Heretic

Also by Ayaan Hirsi Ali

The Caged Virgin

Infidel

Nomad

Heretic

Why Islam Needs a Reformation Now

Ayaan Hirsi Ali

HARPER

NEW YORK • LONDON • TORONTO • SYDNEY

HARPER

A hardcover edition of this book was published in 2015 by HarperCollins Publishers.

FIRST HARPER PAPERBACK EDITION PUBLISHED 2016.

Library of Congress Cataloging-in-Publication Data has been applied for.

ISBN 978-0-06-233394-0 (pbk.)

17 18 19 20 OV/RRD 10 9 8 7 6

To Niall and Thomas

· CONTENTS ·

Heretic

INTRODUCTION

ONE ISLAM, THREE SETS OF MUSLIMS

On ______, a group of ______ heavily armed, black-clad men burst into a ______ in ______, opening fire and killing a total of ______ people. The attackers were filmed shouting "Allahu akbar!"

Speaking at a press conference, President ______ said: "We condemn this criminal act by extremists. Their attempt to justify their violent acts in the name of a religion of peace will not, however, succeed. We also condemn with equal force those who would use this atrocity as a pretext for Islamophobic hate crimes."

As I revised the introduction to this book, four months before its publication, I could of course have written something more specific, like this:

> On January 7, 2015, two heavily armed, black-clad attackers burst into the offices of *Charlie Hebdo* in Paris, opening fire and killing a total of ten people. The attackers were filmed shouting "Allahu akbar!"

But, on reflection, there seemed little reason to pick Paris. Just a few weeks earlier I could equally as well have written this:

> In December 2014, a group of nine heavily armed, black-clad men burst into a school in Peshawar, opening fire and killing a total of 145 people.

Indeed, I could have written a similar sentence about any number of events, from Ottawa, Canada, to Sydney, Australia, to Baga, Nigeria. So instead I decided to leave the place blank and the number of killers and victims blank, too. You, the reader, can simply fill them in with the latest case that happens to be in the news. Or, if you prefer a more historical example, you can try this:

> In September 2001, a group of 19 Islamic terrorists flew hijacked planes into buildings in New York and Washington, D.C., killing 2,996 people.

For more than thirteen years now, I have been making a simple argument in response to such acts of terrorism. My argument is that it is foolish to insist, as our leaders habitually do, that the violent acts of radical Islamists can be divorced from the religious ideals that inspire them. Instead we must acknowledge that they are driven by a political ideology, an ideology embedded in Islam itself, in the holy book of the Qur'an as well as the life and teachings of the Prophet Muhammad contained in the hadith.

Let me make my point in the simplest possible terms: *Islam is not a religion of peace.*

For expressing the idea that Islamic violence is rooted not in social, economic, or political conditions—or even in theological error—but rather in the foundational texts of Islam itself, I have been denounced as a bigot and an "Islamophobe." I have been silenced, shunned, and shamed. In effect, I have been deemed to be a heretic, not just by Muslims—for whom I am already an apostate—but by some Western liberals as well, whose multicultural sensibilities are offended by such "insensitive" pronouncements.

My uncompromising statements on this topic have incited such vehement denunciations that one would think I had committed an act of violence myself. For today, it seems, speaking the truth about Islam is a crime. "Hate speech" is the modern term for heresy. And in the present atmosphere, anything that makes Muslims feel uncomfortable is branded as "hate."

In these pages, it is my intention to make many people—not only Muslims but also Western apologists for Islam—uncomfortable. I am not going to do this by drawing cartoons. Rather, I intend to challenge centuries of religious orthodoxy with ideas and arguments that I am certain will be denounced as heretical. My argument is for nothing less than a Muslim Reformation. Without fundamental alterations to some of Islam's core concepts, I believe, we shall not solve the burning and increasingly global problem of political violence carried out in the name of religion. I intend to speak freely, in the hope that others will debate equally freely with me on what needs to change in Islamic doctrine, rather than seeking to stifle discussion.

Let me illustrate with an anecdote why I believe this book is necessary.

In September 2013, I was flattered to be called by the then-president of Brandeis University, Frederick Lawrence, and offered an honorary degree in social justice, to be conferred at the university's commencement ceremony in May 2014. All seemed well until six months later, when I received another phone call from President Lawrence, this time to inform me that Brandeis was revoking my invitation. I was stunned. I soon learned that an online petition, organized initially by the Council on American Islamic Relations (CAIR) and located at the website change.org, had been circulated by some students and faculty who were offended by my selection.

Accusing me of "hate speech," the change.org petition began by saying that it had "come as a shock to our community due to her extreme Islamophobic beliefs, that Ayaan Hirsi Ali would be receiving an Honorary Degree in Social Justice this year. The selection of Hirsi Ali to receive an honorary degree is a blatant and callous disregard by the administration of not only the Muslim students, but of any student who has experienced pure hate speech. It is a direct violation of Brandeis University's own moral code as well as the rights of Brandeis students."[1] In closing, the petitioners asked: "How can an Administration of a University that prides itself on social justice and acceptance of all make a decision that targets and disrespects it's [*sic*] own students?" My nomination to receive an honorary degree was "hurtful to the Muslim students and the Brandeis community who stand for social justice."[2]

No fewer than eighty-seven members of the Brandeis faculty had also written to express their "shock and dismay" at a few brief snippets of my public statements, mostly drawn from interviews I had given seven years before. I was, they said, a "divisive individual." In particular, I was guilty of suggesting that:

> violence toward girls and women is particular to Islam or the Two-Third World, thereby obscuring such violence in our midst among non-Muslims, including on our own campus [and] . . . the hard work on the ground by committed Muslim feminist and other progressive Muslim activists and scholars, who find support for gender and other equality within the Muslim tradition and are effective at achieving it.[3]

On scrolling down the list of faculty signatories, I was struck by the strange bedfellows I had inadvertently brought together. Professors of "Women's, Gender and Sexuality Studies" lining up with CAIR, an organization subsequently blacklisted as a terrorist organization by the United Arab Emirates? An authority on "Queer/Feminist Narrative Theory" siding with the openly homophobic Islamists?

It is quite true that in February 2007, when I still resided in Holland, I told the London *Evening Standard*: "Violence is inherent in Islam." This was one of three brief, selectively edited quotations to which the Brandeis faculty took exception. What they omitted to mention in their letter was that, less than three years before, my collaborator on a short documentary film, Theo van Gogh, had been murdered in the street in Amsterdam by a young man of Moroccan parentage named Mohammed Bouyeri. First he shot Theo eight times with a handgun. Then he shot him again as Theo, still clinging to life, pleaded for mercy. Then he cut his throat and attempted to decapitate him with a large knife. Finally, using a smaller knife, he stuck a long note to Theo's body.

I wonder how many of my campus critics have read this letter, which was structured in the style of a fatwa, or religious verdict. It began, "In the name of Allah—the Beneficent—the Merciful" and included, along with numerous quotations from the Qur'an, an explicit threat on my life:

> My *Rabb* [master] give us death to give us happiness with martyrdom. *Allahumma Amen* [Oh, Allah, please accept]. Mrs. Hirshi [*sic*] Ali and the rest of you extremist unbelievers. Islam has withstood many enemies and persecutions throughout History. . . . AYAAN HIRSI ALI YOU WILL SELF-DESTRUCT ON ISLAM![4]

On and on it went in the same ranting vein. "Islam will be victorious through the blood of the martyrs. They will spread its light in every dark corner of this earth and it will drive evil with the sword if necessary back into its dark hole. . . . There will be no mercy shown to the purveyors of injustice, only the sword will be lifted against them. No discussions, no demonstrations, no petitions." The note also included this passage, copied directly from the Qur'an: "Be warned that the death that you are trying to prevent will surely find you, afterwards you will be taken back to the All Knowing and He will tell you what you attempted to do" (62:8).

Perhaps those who have risen to the rarefied heights of the Brandeis faculty can devise a way of arguing that no connection exists between Bouyeri's actions and Islam. I can certainly remember Dutch academics claiming that, behind his religious language, Bouyeri's real motivation in wanting to kill me was socioeconomic deprivation or postmodern alienation. To me, however, when a murderer quotes the Qur'an in justification of his crime, we should at least discuss the possibility that he means what he says.

Now, when I assert that Islam is not a religion of peace I do not mean that Islamic belief makes Muslims naturally violent. This is manifestly not the case: there are many millions of peaceful Muslims in the world. What I do say is that the call to violence and the justification for it are explicitly stated in the sacred texts of Islam. Moreover, this theologically sanctioned violence is there to be activated by any number of offenses, including but not limited to apostasy, adultery, blasphemy, and even something as vague as threats to family honor or to the honor of Islam itself.

Yet from the moment I first began to argue that there was an unavoidable connection between the religion I was raised in and the violence of organizations such as Al-Qaeda and the self-styled Islamic State (henceforth IS, though others prefer the acronyms ISIS or ISIL), I have been subjected to a sustained effort to silence my voice.

Death threats are obviously the most troubling form of intimidation. But there have also been other, less violent methods. Muslim organizations such as CAIR have tried to prevent me from speaking freely, particularly on university campuses. Some have argued that because I am not a scholar of Islamic religion, or even a practicing Muslim, I am not a competent authority on the subject. In other venues, select Muslims and Western liberals have accused me of "Islamophobia," a word designed to be equated with anti-Semitism, homophobia, or other prejudices that Western societies have learned to abhor and condemn.

Why are these people impelled to try to silence me, to protest against my public appearances, to stigmatize my views and drive me off the stage with threats of violence and death? It is not because I am ignorant or ill-informed. On the contrary, my views on Islam are based on my knowledge and experience of being a Muslim, of living in Muslim societies—including Mecca itself, the very center of Islamic belief—and on my years of study of Islam as a practitioner, student, and teacher. The real explanation is clear. It is because they cannot actually refute what I am saying. And I am not alone. Shortly after the attack on *Charlie Hebdo*, Asra Nomani, a Muslim reformer, spoke out against what she calls the "honor brigade"—an organized international cabal hell-bent on silencing debate on Islam.[5]

The shameful thing is that this campaign is effective in the West. Western liberals now seem to collude against critical thought and debate. I never cease to be amazed by the fact that non-Muslims who consider themselves liberals—including feminists and advocates of gay rights—are so readily persuaded by these crass means to take the Islamists' side against Muslim and non-Muslim critics.

In the weeks and months that followed, Islam was repeatedly in the news—and not as a religion of peace. On April 14, six days after Brandeis's disinvitation, the violent Islamist group Boko Haram kidnapped 276 schoolgirls in Nigeria. On May 15, in Sudan, a pregnant woman, Meriam Ibrahim, was sentenced to death for the crime of apostasy. On June 29, IS proclaimed its new caliphate in Iraq and Syria. On August 19, the American journalist James Foley was beheaded on video. On September 2, Steven Sotloff, also an American journalist, shared this fate. The man presiding over their executions was clearly identifiable as being British educated, one of between

3,000 and 4,500 European Union citizens who have become jihadists in Iraq and Syria. On September 26, a recent convert to Islam, Alton Nolen, beheaded his co-worker Colleen Hufford at a food-processing plant in Moore, Oklahoma. On October 22, another criminal turned Muslim convert, named Michael Zehaf-Bibeau, ran amok in the Canadian capital, Ottawa, fatally shooting Corporal Nathan Cirillo, who was on sentry duty. And so it has gone on ever since. On December 15, a cleric named Man Haron Monis took eighteen people hostage in a Sydney café; two died in the resulting shootout. Finally, just as I was finishing this book, the staff of the satirical French weekly *Charlie Hebdo* were massacred in Paris. Masked and armed with AK-47 rifles, the Kouachi brothers forced their way into the offices of the magazine and killed the editor, Stéphane Charbonnier, along with nine other employees and a police officer. They killed another police officer in the street. Within hours, their associate Amedy Coulibaly killed four people, all of them Jewish, after seizing control of a kosher store in the east of the city.

In every case, the perpetrators used Islamic language or symbols as they carried out their crimes. To give a single example, during their attack on *Charlie Hebdo*, the Kouachis shouted "*Allahu akbar*" ("God is great") and "the Prophet is avenged." They told a female member of the staff in the offices they would spare her "because you are a woman. We do not kill women. But think about what you are doing. What you are doing is bad. I spare you, and because I spare you, you will read the Qur'an."[6]

If I had needed fresh evidence that violence in the name of Islam was spreading not only across the Middle East and North Africa but also through Western Europe, across the Atlantic and beyond, here it was in lamentable abundance.

After Steven Sotloff's decapitation, Vice President Joe Biden pledged to pursue his killers to the "gates of hell." So outraged was President Barack Obama that he chose to reverse his policy of ending American military intervention in Iraq, ordering air strikes and deploying military personnel as part of an effort to "degrade and ultimately destroy the terrorist group known as ISIL." But the president's statement of September 10, 2014, is worth reading closely for its critical evasions and distortions:

> Now let's make two things clear: ISIL is not "Islamic." No religion condones the killing of innocents. And the vast majority of ISIL's victims have been Muslim. And ISIL is certainly not a state. . . . ISIL is a terrorist organization, pure and simple. And it has no vision other than the slaughter of all who stand in its way.

In short, Islamic State was neither a state nor Islamic. It was "evil." Its members were "unique in their brutality." The campaign against it was like an effort to eradicate "cancer."

After the *Charlie Hebdo* massacre, the White House press secretary went to great lengths to distinguish between "the violent extremist messaging that ISIL and other extremist organizations are using to try to radicalize individuals around the globe" and a "peaceful religion." The administration, he said, had "enjoyed significant success in enlisting leaders in the Muslim community . . . to be clear about what the tenets of Islam actually are." The very phrase "radical Islam" was no longer to be uttered.

But what if this entire premise is wrong? For it is not just Al-Qaeda and IS that show the violent face of Islamic faith and practice. It is Pakistan, where any statement critical of the Prophet or Islam is labeled as blasphemy and punishable by death. It is Saudi Arabia, where churches and synagogues are outlawed, and where beheadings are a legitimate form of punishment, so much so that there was almost a beheading a day in August 2014. It is Iran, where stoning is an acceptable punishment and homosexuals are hanged for their "crime." It is Brunei, where the sultan is reinstituting Islamic sharia law, again making homosexuality punishable by death.

We have now had almost a decade and a half of policies and pronouncements based on the assumption that terrorism or extremism can and must be differentiated from Islam. Again and again in the wake of terrorist attacks around the globe, Western leaders have hastened to declare that the problem has nothing to do with Islam itself. For Islam is a religion of peace.

These efforts are well meaning, but they arise from a misguided conviction, held by many Western liberals, that retaliation against Muslims is more to be feared than Islamist violence itself. Thus, those responsible for the 9/11 attacks were represented not as Muslims but as terrorists; we focused on their tactics rather than on the ideology that justified their horrific acts. In the process, we embraced those "moderate" Muslims who blandly told us Islam was a religion of peace and marginalized dissident Muslims who were attempting to pursue real reform.

Today, we are still trying to argue that the violence is the work of a lunatic fringe of extremists. We employ medical metaphors, trying to define the phenomenon as some kind of foreign body alien to the religious milieu in which it flourishes. And we make believe that there are extremists just as bad as the jihadists in our own midst. The president of the United States even went so far as to declare, in a speech to the

United Nations General Assembly in 2012: "The future must not belong to those who slander the Prophet of Islam"—as opposed, presumably, to those who go around killing the slanderers.

Some people will doubtless complain that this book slanders Muhammad. But its aim is not to give gratuitous offense, but to show that this kind of approach wholly—not just partly, but wholly—misunderstands the problem of Islam in the twenty-first century. Indeed, this approach also misunderstands the nature and meaning of liberalism.

For the fundamental problem is that the majority of otherwise peaceful and law-abiding Muslims are unwilling to acknowledge, much less to repudiate, the theological warrant for intolerance and violence embedded in their own religious texts.

It simply will not do for Muslims to claim that their religion has been "hijacked" by extremists. The killers of IS and Boko Haram cite the same religious texts that every other Muslim in the world considers sacrosanct. And instead of letting them off the hook with bland clichés about Islam as a religion of peace, we in the West need to challenge and debate the very substance of Islamic thought and practice. We need to hold Islam accountable for the acts of its most violent adherents and demand that it reform or disavow the key beliefs that are used to justify those acts.

At the same time, we need to stand up for our own principles as liberals. Specifically, we need to say to offended Western Muslims (and their liberal supporters) that it is not we who must accommodate their beliefs and sensitivities. Rather, it is they who must learn to live with our commitment to free speech.

Three Sets of Muslims

Before we begin to speak about Islam, we must understand what it is and recognize certain distinctions within the Muslim world. The distinctions I have in mind are not the conventional ones among Sunni, Shia, and other branches of the faith. Rather, they are broad sociological groupings defined by the nature of their observance. I will subdivide Muslims. I will not subdivide Islam.

Islam is a single core creed based on the Qur'an, the words revealed by the Angel Gabriel to the Prophet Muhammad, and the hadith, the accompanying works that detail Muhammad's life and words. Despite some sectarian differences, this creed unites all Muslims. All, without exception, know by heart these words: "I bear witness that there is no God but Allah; and Muhammad is His messenger." This is the Shahada, the Muslim profession of faith.

The Shahada may seem a declaration of belief no different from any other to Westerners used to individual freedom of conscience and religion. But the reality is that the Shahada is both a religious *and* a political symbol.

In the early days of Islam, when Muhammad was going from door to door trying to persuade the polytheists to abandon their idols of worship, he was *inviting* them to accept that there was no god but Allah and that he was Allah's messenger, much as Christ had asked the Jews to accept that he was the son of God. However, after ten years of trying this kind of persuasion, Muhammad and his small band of believers went to Medina and from that moment Muhammad's mission took on a political dimension. Unbelievers were still invited to submit to Allah, but, after Medina, they were attacked if they refused. If defeated, they were given the option either to convert or to die. (Jews and Christians could retain their faith if they submitted to paying a special tax.)

No symbol represents the soul of Islam more than the Shahada. But today there is a contest within Islam for the ownership of that symbol. Who owns the Shahada? Is it those Muslims who want to emphasize Muhammad's years in Mecca, or those who are inspired by his conquests after Medina? There are millions upon millions of Muslims who identify themselves with the former. Increasingly, however, they are challenged by fellow believers who want to revive and reenact the political version of Islam born in Medina—the version that took Muhammad from being a wanderer in the desert to a symbol of absolute morality.

On this basis, I believe we can distinguish three different groups of Muslims.

The first group is the most problematic. These are the fundamentalists who, when they say the Shahada, mean: "We must live by the strict letter of our creed." They envision a regime based on sharia, Islamic religious law. They argue for an Islam largely or completely unchanged from its original seventh-century version. What is more, they take it as a requirement of their faith that they impose it on everyone else.

I was tempted to call this group "Millenarian Muslims," because their fanaticism is reminiscent of the various fundamentalist sects that flourished in medieval Christendom prior to the Reformation, most of which combined fanaticism and violence with anticipation of the end of the world.[7] But the analogy is imperfect. Whereas Shiite doctrine looks forward to the return of the Twelfth Imam and the global triumph of

Islam, Sunni zealots are more likely to aspire to the forcible creation of a new caliphate here on earth. Instead, then, I shall call them Medina Muslims, in that they see the *forcible* imposition of sharia as their religious duty. They aim not just to obey Muhammad's teaching, but also to emulate his warlike conduct after his move to Medina. Even if they do not themselves engage in violence, they do not hesitate to condone it.

It is Medina Muslims who call Jews and Christians "pigs and monkeys" and preach that both faiths are, in the words of the Council on Foreign Relations Fellow (and former Islamist) Ed Husain, "false religions." It is Medina Muslims who prescribe beheading for the crime of "nonbelief" in Islam, death by stoning for adultery, and hanging for homosexuality. It is Medina Muslims who put women in burqas and beat them if they leave their homes alone or if they are improperly veiled. It was Medina Muslims who in July 2014 went on a rampage in Gujranwala, Pakistan, setting eight homes on fire and killing a grandmother and her two granddaughters, all because of the posting of an allegedly blasphemous photo on an eighteen-year-old's Facebook page.

Medina Muslims believe that the murder of an infidel is an imperative if he refuses to convert voluntarily to Islam. They preach jihad and glorify death through martyrdom. The men and women who join groups such as Al-Qaeda, IS, Boko Haram, and Al-Shabaab in my native Somalia—to name just four of hundreds of jihadist organizations—are all Medina Muslims.

Are the Medina Muslims a minority? Ed Husain estimates that only 3 percent of the world's Muslims understand Islam in these militant terms. But out of well over 1.6 billion believers, or 23 percent of the globe's population, that 48 million seems to be more than enough. Based on survey data on attitudes toward sharia in Muslim countries, I would put the proportion significantly higher;[8] I also believe it is rising as Muslims and converts to Islam gravitate toward Medina. Either way, Muslims who belong to this group are not open to persuasion or engagement by either Western liberals or Muslim reformers. They are not the intended audience for this book. They are the reason for writing it.

The second group—and the clear majority throughout the Muslim world—consists of Muslims who are loyal to the core creed and worship devoutly but are not inclined to practice violence. I call them Mecca Muslims. Like devout Christians or Jews who attend religious services every day and abide by religious rules in what they eat and wear, Mecca Muslims focus on religious observance. I was raised a Mecca Muslim. So were the majority of Muslims from Casablanca to Jakarta.

Yet the Mecca Muslims have a problem: their religious beliefs exist in an uneasy tension with modernity—the complex of economic, cultural, and political innovations that not only reshaped the Western world but also dramatically transformed the developing world as the West exported it. The rational, secular, and individualistic values of modernity are fundamentally corrosive of traditional societies, especially hierarchies based on gender, age, and inherited status.

In Muslim-majority countries, the power of modernity to transform economic, social, and (ultimately) power relations can be limited. Muslims in these societies can use cell phones and computers without necessarily seeing a conflict between their religious faith and the rationalist, secular mindset that made modern technology possible. In the West, however, where Islam is a minority religion, devout Muslims live in what is best described as a state of cognitive dissonance. Trapped between two worlds of belief and experience, these

Muslims are engaged in a daily struggle to adhere to Islam in the context of a secular and pluralistic society that challenges their values and beliefs at every turn. Many are able to resolve this tension only by withdrawing into self-enclosed (and increasingly self-governing) enclaves. This is called cocooning, a practice whereby Muslim immigrants attempt to wall off outside influences, permitting only an Islamic education for their children and disengaging from the wider non-Muslim community.[9]

To many such Muslims, after years of dissonance, there appear to be only two alternatives: either leave Islam altogether, as I did, or abandon the dull routine of daily observance for the uncompromising Islamist creed offered by those—the Medina Muslims—who explicitly reject the West's modernity.

It is my hope to engage this second group of Muslims—those closer to Mecca than Medina—in a dialogue about the meaning and practice of their faith. I hope that they will be one of the primary audiences for this book.

Of course, I recognize that these Muslims are not likely to heed a call for doctrinal reformation from someone they regard as an apostate and infidel. But they may reconsider if I can persuade them to think of me not as an apostate, but as a heretic: one of a growing number of people born into Islam who have sought to think critically about the faith we were raised in. It is with this third group—only a few of whom have left Islam altogether—that I would now identify myself.

These are the Muslim dissidents; call them the Modifying Muslims. A few of us have been forced by experience to conclude that we could not continue to be believers; yet we remain deeply engaged in the debate about Islam's future. The majority of dissidents are reforming believers—among them clerics who have come to realize that their religion must change if its followers are not be condemned to an interminable cycle of political violence.

I shall have more to say in what follows about this neglected—indeed largely unknown—group. For now, it is enough to say that I choose to identify myself with the dissidents. In the eyes of the Medina Muslims, we are all heretics, because we have had the temerity to challenge the applicability of seventh-century teachings to the twenty-first-century world.

The dissidents include people such as Abd Al-Hamid Al-Ansari, the former dean of Islamic law at Qatar University, who disavows the hatred of religions other than Islam. He has quoted at length a Saudi woman who asked why her daughter should be taught to hate non-Muslims: "Do they expect me to hate the Jewish scientist who discovered insulin, which I use to treat my mother? Am I supposed to teach my daughter that she should hate Edison, who invented the lightbulb, which lights up the Islamic world? Should I hate the scientist who discovered the cure for malaria? Should I teach my daughter to hate people merely because their religion is different? Why do we turn our religion into a religion of hatred toward those who differ from us?" Al-Ansari then quotes a response by a leading Saudi cleric, who replied, "This is none of your business" and "cooperation with the infidels is permitted, but only as a reward for services, and not out of love." Al-Ansari's plea is to "make religious discourse more human."

And that is precisely the thing Western-based reformers such as Irshad Manji, Maajid Nawaz, and Zuhdi Jasser are seeking: what they have in common is an attempt to modify, adapt, and reinterpret Islamic practice in order to *make religious discourse more human.* (For further details on the Modifying Muslims, see the Appendix.)

How many Muslims belong to each group? Even if it were possible to answer that question definitively, I am not sure that it matters. On the airwaves, over social media, in far too many mosques, and of course on the battlefield, the Medina Muslims have captured the world's attention. Most disturbing, the number of Western-born Muslim jihadists is sharply increasing. The UN estimated in November 2014 that some 15,000 foreign fighters from at least eighty nations have traveled to Syria to join the radical jihadists.[10] Roughly a quarter of them come from Western Europe. And it is not just young men. Between 10 and 15 percent of those traveling to Syria from some Western countries are female, according to estimates from the ICSR research group.[11]

But there are more troubling statistics. According to estimates by the Pew Research Center, the Muslim population of the United States is set to increase from around 2.6 million today to 6.2 million in 2030, mainly as a result of immigration, as well as above-average fertility. Although in relative terms this will still represent less than 2 percent of the total U.S. population (1.7 percent, to be precise, compared with around 0.8 percent today), in absolute terms that will be a larger population than in any West European country except France.[12]

As an immigrant of Somali origin, I have no objection whatever to millions of other people from the Muslim world coming to America to seek a better life for themselves and their families. My concern is with the attitudes many of these new Muslim Americans will bring with them (see table 1).

Approximately two fifths of Muslim immigrants between now and 2030 will be from just three countries: Pakistan, Bangladesh, and Iraq. Another Pew study—of opinion in the Muslim world—shows just how many people in these countries hold views that most Westerners would regard as extreme.[13] Three quarters of Pakistanis and more than two fifths of Bangladeshis and Iraqis think that those who leave Islam should suffer the death penalty. More than 80 percent of Pakistanis and two thirds of Bangladeshis and Iraqis regard sharia law as the revealed word of God. Similar proportions say that Western entertainment hurts morality. Only tiny fractions would be comfortable if their daughters married Christians. Only minorities regard honor killings of women as never justified. A quarter of Bangladeshis and one in eight Pakistanis think that suicide bombings in defense of Islam are often or sometimes justified.

Medina Muslims can exploit views such as these to pose a threat to us all. In the Middle East and elsewhere, their vision of a violent return to the days of the Prophet potentially spells death for hundreds of thousands and subjugation for millions. In the West, it implies not only an increasing risk of terrorism but also a subtle erosion of the hard-won achievements of feminists and campaigners for minority rights.

Medina Muslims are also undermining the position of those Mecca Muslims attempting to lead a quiet life in their cultural cocoons throughout the Western world. Yet those under the greatest threat are the dissidents and reformers: the Modifying Muslims. They are the ones who face ostracism and rejection, who must brave all manner of insults, who must deal with the death threats—or face death itself. So far, their efforts have been diffuse and individual, compared with the highly organized collective action of the Medina Muslims. We owe it to the dissidents—to their courage and their convictions—to change that.

Indeed, I have come to the conclusion that the only viable strategy that can hope to contain the threat posed by the

ATTITUDES IN MAJORITY MUSLIM COUNTRIES WITH LARGE CURRENT AND PROJECTED MIGRATION TO THE UNITED STATES [14]

Percentage of Muslims who . . .	Pakistan	Bangladesh	Iraq
Favor the death penalty for leaving Islam	75	43	41
Say it's necessary to believe in God to be moral	85	89	91
Agree that converting others is a religious duty	85	69	66
Say sharia is revealed word of God	81	65	69
Say religious leaders should have some or large influence	54	69	57
Say Western entertainment hurts morality	88	75	75
Say polygamy is morally acceptable	37	32	46
Say honor killings are never justified when female committed the offense	45	34	22
Say suicide bombing in defense of Islam is often or sometimes justified	13	26	7
Say a wife should be able to divorce her husband	26	62	14
Say they would be very/somewhat comfortable with their daughter marrying a Christian	3	10	4

Medina Muslims is to side with the dissidents and reformists and to help them a) identify and repudiate those parts of Muhammad's moral legacy that stem from Medina and b) persuade the Mecca Muslims to accept this change and reject the Medina Muslims' summons to intolerance and war.

This book is not a work of history. I do not offer a new explanation for the fact that more and more Muslims have embraced the most violent elements of Islam in my lifetime—why, in short, the Medina Muslims are in the ascendant today. I do seek to challenge the view, almost universal among Western liberals, that the explanation lies in the economic and political problems of the Muslim world and that these, in turn, can be explained in terms of Western foreign policy. This is to attach too much importance to exogenous forces. There are other parts of the world that have struggled to make democracy work or to cope with oil wealth. There are other peoples besides Muslims who have complaints about U.S. "imperialism." Yet there is precious little evidence of an upsurge in terrorism, suicide bombings, sectarian warfare, medieval punishments, and honor killings in the non-Muslim world. There is a reason why an increasing proportion of organized violence in the world is happening in countries where Islam is the religion of a substantial share of the population.

The argument in this book is that *religious doctrines matter and are in need of reform.* Non-doctrinal factors—such as the Saudis' use of oil revenues to fund Wahhabism and Western support for the Saudi regime—are important, but *religious doctrine is more important.* Hard as it may be for many Western academics to believe, when people commit violent acts in the name of religion, they are not trying somehow to dignify their underlying socioeconomic or political grievances.

Islam is at a crossroads. Muslims, not by the tens or hundreds but by the tens of millions and eventually hundreds of millions, need to make a conscious decision to confront, debate, and ultimately reject the violent elements within their religion. To some extent—not least because of widespread revulsion at the unspeakable atrocities of IS, Al-Qaeda, and the rest—this process has already begun. But ultimately it needs leadership from the dissidents. And they in turn stand no chance without support from the West.

Imagine if, in the Cold War, the West had lent its support not to the dissidents in Eastern Europe—to the likes of Václav Havel and Lech Wałęsa—but to the Soviet Union, as the representative of "moderate Communists," in the hope that the Kremlin would give us a hand against terrorists such as the Red Army Faction. Imagine if a "Manchurian candidate" president had told the world: "Communism is an ideology of peace."

That would have been disastrous. Yet that is essentially the West's posture toward the Muslim world today. We ignore the dissidents. Indeed, we do not even know their names. We delude ourselves that our deadliest foes are somehow not actuated by the ideology they openly affirm. And we pin our hopes on a majority that is conspicuously without any credible leadership, and indeed shows more sign of being susceptible to the arguments of the fanatics than to those of the dissidents.

Five Amendments

Not everyone will accept this argument, I know. All I ask of those who do not is that they defend my right to make it. But for those who do accept the proposition that Islamic extremism is rooted in Islam, the central question is: What needs to happen for us to defeat the extremists for good? Economic, political, judicial, and military tools have been proposed and some of them deployed. But I believe these will have little effect unless Islam itself is reformed.

Such a Reformation has been called for repeatedly—by Muslim activists such as Muhammad Taha and Western scholars such as Bernard Lewis—at least since the fall of the Ottoman Empire and the subsequent abolition of the Caliphate. In that sense, this is not an original work. What is original is that I specify precisely what needs to be reformed. I have identified five precepts central to the faith that have made it resistant to historical change and adaptation. Only when these five things are recognized as inherently harmful and when they are repudiated and nullified will a true Muslim Reformation have been achieved. The five things to be reformed are:

1. Muhammad's semi-divine and infallible status along with the literalist reading of the Qur'an, particularly those parts that were revealed in Medina;
2. The investment in life after death instead of life before death;
3. Sharia, the body of legislation derived from the Qur'an, the hadith, and the rest of Islamic jurisprudence;
4. The practice of empowering individuals to enforce Islamic law by commanding right and forbidding wrong;
5. The imperative to wage jihad, or holy war.

All these tenets must be either reformed or discarded. In the chapters that follow I shall discuss each of them and make the case for their reformation.

I recognize that such an argument is going to make many Muslims uncomfortable. Some are bound to say that they are offended by my proposed amendments. Others will no doubt contend that I am not qualified to discuss these complex issues of theological and legal tradition. I am also afraid—genuinely afraid—that it will make a few Muslims even more eager to silence me.

But this is not a work of theology. It is more in the nature of a public intervention in the debate about the future of Islam. The biggest obstacle to change within the Muslim world is precisely its suppression of the sort of critical thinking I am attempting here. If nothing else comes of it, I will consider this book a success if it helps to spark a serious discussion of these issues among Muslims themselves. That, in my opinion, would represent a first step, however hesitant, toward the Reformation that Islam so desperately needs.

For their part, many Westerners may be inclined to dismiss these propositions as quixotic. Other religions have undergone a process of reform, modifying core beliefs and adopting more tolerant and flexible attitudes compatible with modern, pluralistic societies. But what hope can there be to reform a religion that has resisted change for 1,400 years? If anything, Islam today seems, from the Western point of view, to be moving backward, not forward. Ironically, this book is written at a time when many in the West have begun to despair of winning the struggle against Islamic extremism, and when the hopes associated with the so-called Arab Spring have largely proved to be illusory.

I agree that the Arab Spring was an illusion, at least in terms of Western expectations. From the outset, I regarded parallels with the Prague Spring of 1968 or the Velvet Revolution of 1989 as facile and doomed to disappointment. Nevertheless, I think many Western observers have missed the underlying import of the Arab Spring. Something was—and still is—definitely afoot within the Muslim world. There is a genuine constituency for change that was never there before. It is a constituency, I shall argue, that we overlook at our peril.

In short, this is an optimistic book, a book that seeks to inspire not another war on terror or extremism but rather a real debate within and about the Muslim world. It is a book that attempts to explain what elements such a Reformation might change, written from the perspective of someone who has been at various times all three kinds of Muslim: a cocooned believer, a fundamentalist, and a dissident. My journey has gone from Mecca to Medina to Manhattan, and to the idea of a Modified Islam.

The absence of a Muslim Reformation is what ultimately drove me to become an infidel, a nomad, and now a heretic. Future generations of Muslims deserve better, safer options. Muslims should be able to welcome modernity, not be forced to wall themselves off, or live in a state of cognitive dissonance, or lash out in violent rejection.

The Muslim world is currently engaged in a massive struggle to come to terms with the challenge of modernity. The Arab Spring and Islamic State are just two versions of the reaction to that challenge. We in the West must not limit ourselves solely to military means in order to defeat the jihadists. Nor can we hope to cut ourselves off from contact with them. For these reasons, we have an enormous stake in how the struggle over Islam plays out. We cannot remain on the sidelines as though the outcome has nothing to do with us. If the Medina Muslims win and the hope for a Muslim Reformation dies, the rest of the world will pay an enormous price. And, with all the freedoms we take for granted, Westerners may have the most to lose.

That is why I am also addressing this book to Western liberals—not just to those who saw fit to disinvite me from Brandeis but also to all the many others who would have done the same if their university had offered me an honorary degree.

You who call yourselves liberals must understand that it is your way of life that is under threat. Withdraw my right to speak freely, and you jeopardize your own in the future. Ally yourselves with the Islamists at your peril. Tolerate their intolerance at your peril.

In all kinds of ways, feminists and gay rights activists offer their support to Muslim women and gays in the West and, increasingly, in Muslim-majority countries. However, most shy away from linking the abuses they are against—from child marriage to the persecution of homosexuals—to the religious tenets on which the abuses are based. To give just a single example, in August 2014 the theocratic regime in Tehran executed two men, Abdullah Ghavami Chahzanjiru and Salman Ghanbari Chahzanjiri, apparently for violating the Islamic Republic's law against sodomy. That law is based on the Qur'an and the hadith.

People like me—some of us apostates, most of us dissident Muslims—need your support, not your antagonism. We who have known what it is to live without freedom watch with incredulity as you who call yourselves liberals—who claim to believe so fervently in individual liberty and minority rights—make common cause with the forces in the world that manifestly pose the greatest threats to that very freedom and those very minorities.

I am now one of you: a Westerner. I share with you the pleasures of the seminar rooms and the campus cafés. I know we Western intellectuals cannot lead a Muslim Reformation. But we do have an important role to play. We must no longer accept limitations on criticism of Islam. We must reject the notions that only Muslims can speak about Islam, and that any critical examination of Islam is inherently "racist." Instead of contorting Western intellectual traditions so as not to offend our Muslim fellow citizens, we need to defend the Muslim dissidents who are risking their lives to promote the human rights we take for granted: equality for women, tolerance of all religions and orientations, our hard-won freedoms of speech and thought. We support the women in Saudi Arabia who wish to drive, the women in Egypt who are protesting against sexual assault, the homosexuals in Iraq, Iran, and Pakistan, the young Muslim men who want not martyrdom but the freedom to leave their faith. But our support would be more effective if we acknowledged the theological bases of their oppression.

In short, we who have the luxury of living in the West have an obligation to stand up for liberal principles. Multiculturalism should not mean that we tolerate another culture's intolerance. If we do in fact support diversity, women's rights, and gay rights, then we cannot in good conscience give Islam a free pass on the grounds of multicultural sensitivity. And we need to say unambiguously to Muslims living in the West: If you want to live in our societies, to share in their material benefits, then you need to accept that our freedoms are not optional. They are the foundation of our way of life; of our civilization—a civilization that learned, slowly and painfully, not to burn heretics, but to honor them.

Indeed, one highly desirable outcome of a Muslim Reformation would be to redefine the meaning of the word "heretic" itself. Religious reformations always shift the meaning of this term: today's heretic becomes tomorrow's reformer, while today's defender of religious orthodoxy becomes tomorrow's Torquemada. A Muslim Reformation would have the happy effect of turning the tables on those I am threatened by—rendering them the heretics, not me.

CHAPTER 1

THE STORY OF A HERETIC

My Journey Away from Islam

I was raised a practicing Muslim and remained one for almost half my life. I attended madrassas and memorized large parts of the Qur'an. As a child, I lived in Mecca for a time and frequently visited the Grand Mosque. As a teenager, I joined the Muslim Brotherhood. In short, I am old enough to have seen Islam's bifurcation in the latter half of the twentieth century between the everyday faith of my parents and the intolerant, militant jihadism preached by the people I call the Medina Muslims. So let me begin with the Islam in which I grew up.

I was about three years old when my grandmother started teaching me what little she had memorized of the Qur'an under the feathery leaves of the Somali *talal* tree. She could not read or write—literacy began to be promoted in Somalia only in 1969, the year I was born—and had no concept of Arabic. Instead, she worshipped the book, picking it up with great reverence, kissing it and placing it on her forehead before carefully and gently laying it back down. We could not touch the Qur'an without first washing our hands. My mother was the same way, except she had memorized a bit more and spoke a little Arabic. She had learned the prayers by heart and could also recite fearsome incantations, warning me that I would burn in hellfire for any misdeeds.

My mother was born under a tree and grew up in the desert, and she was a wanderer when she was young, making it as far as Aden in Yemen, across the Red Sea. She was subjected to an arranged marriage and sent to Kuwait with her husband. As soon as her own father died, she divorced this husband. She met my father through her older sister when he was teaching people in the Somali capital how to read and write. My mother was one of his best students, with a quick and clever way with words. My father already had a wife, so my mother became his second. My father was a political man, an opposition leader trying to change Somalia, which was then ruled by the dictator Siad Barre. When I was two, the authorities came for him and took him away to the old Italian prison, otherwise known as "the Hole." So, for most of my early years, it was simply my mother, my brother, my sister, my grandmother, and me.

My first real school was a religious *dugsi*—a shed offering shelter from the burning sun. Between thirty and forty children sat under a roof held up by poles, surrounded by a thicket of trees. We had the only spot of shade. At the front and center of the space was a foot-high wooden table on which rested a large copy of the Qur'an. Our teacher wore the traditional Somali man's garb of a sarong and a shirt, and he made us chant the verses, much as American and European preschool students learn to chant short poems and nursery rhymes. If we forgot or we were simply not loud enough or our voices dipped too low, he would take his stick and prod or whack us.

We chanted again when students misbehaved. If you were disobedient, if you failed to learn what you were supposed to have learned, you were sent to the middle of the shed. The worst offender was hoisted high in a hammock and swung back and forth in the air. The rest of us were given little sticks and we raised our sticks above our heads and stood underneath, hitting the disobedient child through the open holes of the hammock, calling out verses from the Qur'an, chanting about the Day of Judgment, when the sun goes black and the hellfires burn.

Every punishment at school or at home seemed to be laced with threats of hellfire and pleas for death or destruction: may you suffer this disease or that, and may you burn in hell. And yet in the evening, when the sun had dropped below the horizon and the cool night air reigned over us, my mother would face toward Mecca and say the evening prayer. Again and again, three maybe four times, she would recite the words, the opening verses of the Qur'an, and other verses, moving from standing with her hand across her womb, to bowing down, to prostrating herself, to sitting, then prostrating, then sitting again. There was an entire ritual of words and movement, and it repeated itself each night.

After her prayers, we sat with cupped hands under the *talal* tree, begging Allah to release my father from prison. These were supplications to God to make life easy, asking Allah to be patient with us, to give us resilience, to convey upon us forgiveness and peace. "I seek shelter in Allah," she would chant. "Allah the most merciful, the most kind . . . My Lord, forgive me, have mercy upon me, guide me, give me health and grant me sustenance and exalt me and set right my affairs." It became as familiar and soothing as a lullaby, as far removed as could be imagined from the clashing sticks and taunting words of the *dugsi*.

The supplications seemed to work. Thanks to the help of a relative, my father was able to escape from jail and flee to Ethiopia. The obvious thing would have been for my mother to take us to Ethiopia, too. But my mother would not go to Ethiopia. Because it was predominantly Christian, to her it was nothing but a sea of infidels in an unclean land. She preferred to go to Saudi Arabia, the cradle of Islam, seat of its holiest places, Mecca and Medina. So she got a false passport and airline tickets, and then, one morning when I was eight years old, my grandmother woke us before dawn, dressed us in our good clothes, and by the time the day was over, we were in Saudi Arabia.

We settled in Mecca, the spiritual heart of Islam, the place to which nearly every Muslim dreams of making a pilgrimage once in his or her life. We could enact that pilgrimage every week by taking the bus from our apartment to the Grand Mosque. At eight years old, I had already performed the Umra, the little version of the full pilgrimage to Mecca, the Hajj, the fifth pillar of the Muslim faith, which washes away the pilgrim's sins. Now, moreover, we could study Islam as it was taught in Saudi religious schools, rather than in a Somali shed. My sister, Haweya, and I were enrolled in a Qur'an school for girls; my brother, Mahad, went to a madrassa for boys. Previously I had been taught that all Muslims were united in brotherhood, but here I discovered that the brotherhood of Muslims did not preclude racial and cultural prejudice. What we had learned of the Qur'an in Somalia was not good enough for the Saudis. We did not know enough; we mumbled instead of reciting. We did not learn to write any of the passages, we just learned to memorize each verse, repeating it slowly again and again. The Saudi girls were light-skinned and called us *abid*, or slaves—in fact, the Saudis had legally abolished slavery just five years before I was born. At home, my mother now made us pray five times each day, performing the rituals of washing and robing each time.

It was here that I encountered for the first time the strict application of sharia law. In the public squares, every Friday, after the ritual prayers, men were beheaded or flogged, women were stoned, and thieves had their hands cut off amid great spurts of blood. The rhythm of chanted prayers was replaced by the reverberation of metal blades slicing through flesh and hitting stone. My brother—who, unlike me, was allowed to witness these punishments—used the nickname "Chop-Chop Square" for the one closest to us. We never questioned the ferocity of the punishments. To us, it was simply more hellfire.

But the Grand Mosque, with its high columns, elaborate tiles, and polished floors, was more beguiling. Here, in the cool shade, my mother could walk seven times around the Kaaba, the holy building at the center of the mosque. This tranquillity was interrupted only in the month of the Hajj, the Islamic ritual pilgrimage, when we could not leave our apartment for fear of being trampled by the masses of believers streaming down the streets, and when even the simplest conversations had to be shouted over the din of constant prayer.

It was in Mecca that I first became conscious of the differences between my father's vision of Islam and my mother's. After my father came from Ethiopia to join us, he insisted that we pray not separated by sex in separate rooms of the apartment, as was Saudi tradition, but together as a family. He did not throw the specter of hell in our faces, and once a week he taught us the Qur'an, reading from it and trying to translate it, infusing it with his own interpretations. He would tell me and my brother and my sister that God hadn't put us on earth to punish us; He had put us on earth to worship him. I would look up and nod, but then, the next morning or afternoon, if I disobeyed my mother, she would once again revert to hellfire and eternal punishment.

After a time, we moved to Riyadh, where my father was working as a translator of Morse code for a government ministry. We had a house with a men's side and a women's, although unlike our neighbors, the five of us moved easily between the two sides. My father did not behave like the Saudi men. He did not do the shopping or handle all the outside transactions. Moreover, he continued to absent himself, returning to Ethiopia, where the Somali opposition was based. The neighbors openly pitied my mother for having to go out of the house alone. In turn, my mother looked down on the Saudi girls for teaching Haweya and me the rudiments of belly-dancing. She wanted us to live only according to "pure Islam," which to her meant no singing or dancing, no laughter or joy.

A little over a year later, when I was nine, we left as quickly as we came. My father was deported by the Saudi government. The reasons were unclear to me, but they no doubt related to his ongoing Somali opposition activities. We had twenty-four hours to pack and fly—this time to Ethiopia. After a year and a half there, my mother's antipathy to the country necessitated yet another move: to Kenya.

In Nairobi, Haweya and I went to school. English was not the only thing I learned there. I soon discovered that I did not know the most basic things, like the date and how to tell time. Ethiopia had a sidereal calendar; Saudi Arabia used an Islamic lunar calendar; in Somalia, my grandmother told time solely by the sun and her year consisted of ten months. It was only as a ten-year-old in Kenya that I learned it was the year 1980. For the Saudis, it was the Islamic year 1400; for the Ethiopians, by their way of reckoning, it was still 1978.

My mother nevertheless remained steadfast in her faith: she refused to believe that the things we were taught in school, such as the moon landings and evolution, were true; Kenyans might be descended from apes, but not us: she made us recite our bloodline to prove the point. As soon as I turned fourteen, she enrolled me in the Muslim Girls' Secondary School on Park Road so that my sister and I could have a more modest uniform. Now we could wear trousers underneath our skirts. Our heads could be covered in white headscarves. At least, those things were permitted. But at that time few girls complied.

I Embrace the Islam of Medina

Then, when I was sixteen, I discovered a way of being a better Muslim. A new teacher arrived to teach us religious education. Sister Aziza was a Sunni Muslim from the Kenyan coast who had converted to Shia Islam following her marriage. She wore the full hijab; almost nothing was visible except her face. She even wore gloves and socks to keep her fingers and toes concealed.

Previously we had been taught Islam as history: dates and caliphates. Aziza did not teach; she preached. Better, she seemed to reason with us, questioning us, leading us. "What makes you different from the infidels?" The correct answer was the Shahada, the Muslim's profession of faith. "How many times a day should you pray?" We knew that the answer was five. "How many times did you pray yesterday?" We looked nervously at one another.

This was a far more seductive method of teaching than any stick, and Sister Aziza did not care how long it took. As she liked to say: "This is how Allah and the Prophet want you to dress. But you should only do it when you are ready," adding, "When you're ready for it, you'll choose, and then you'll never take it off."

Another novelty: Aziza did not read the Qur'an in Arabic, but from English translations, and unlike my previous teachers—including my mother—she said she was not forcing us. She was simply sharing with us Allah's words, His wishes, His desires. If we chose not to please Allah, then of course we would burn in hell. But if we pleased Him, then we would go to paradise.

There was an element of choice here that was irresistible. Our parents, and certainly my mother, could never be pleased, whatever we did. Our earthly lives could not be changed. In a few years or less, we would find ourselves extracted from school, sent off into arranged marriages. We seemed to have no choices. But our *spiritual* lives were another matter. Those lives could be transformed, and Sister Aziza could show us the way. And then we, in turn, could show others the way. It is hard to overstate how empowering this message was.

It took me a while, but when I embraced Sister Aziza's path, I did it in earnest. I prayed without fail five times a day. I went to a tailor to buy a vast, voluminous cloak that clinched tight around my wrists and billowed down to my toes. I wore it over my school uniform and wound a black scarf over my hair and shoulders. I put it on in the morning to walk to school and again before I left the school gates to return home. As I walked along the streets, covered, I had to move very deliberately because it was easy to trip over the billowing fabric. It was hot and cumbersome. In those moments, as my giant black figure moved slowly down the street, my mother was finally happy with me. But I was not doing it for her. I was doing it for Allah.

Sister Aziza was not the only new kind of Muslim I encountered at that time. There were now preachers going from door to door, like the self-appointed imam Boqol Sawm. His name meant "He Who Fasts for a Hundred Days," and in person he more than lived up to his name. He was so thin that he looked like skin stretched over bone. While Sister Aziza wore the hijab, Boqol Sawm wore a Saudi robe, a bit short, so that it showed his bony ankles. It seemed he did nothing except walk around Old Racecourse Road, our neighborhood in Nairobi, knocking on doors, sermonizing, and leaving cassette tapes for the women who invited him in. There were no Electrolux salesmen with their vacuums going door to door in Old Racecourse Road, just Boqol Sawm and his sermons. He would sometimes come inside, too, as long as there was a curtain to separate him from the women, who listened to the cassettes he left behind and traded them. They played the sermons while they were washing and cooking. Gradually they stopped wearing colorful clothes and shrouded themselves in the *jilbab*, a long, loose-fitting coat, and wrapped scarves around their heads and necks.

If Aziza's methods of indoctrination were subtle, Boqol Sawm favored the more familiar verbal bludgeoning I had first encountered back in Somalia. He shouted his verses in Arabic and Somali and highlighted what was forbidden and what was permitted—in a manner so strident that he got himself shut out of the local mosque. Women, he preached, should be available to men at any time, "even on the saddle of a camel," except during the days of the month when they were unclean. This might not seem a very appealing message for a female audience, yet for many women he was mesmerizing. And for their sons, he was positively transforming.

More and more Somali teenage boys in our expatriate community had started hanging out in the street in proto-gangs, dropping out of school, chewing qat, committing petty crimes, harassing and even raping women, spinning completely beyond their mothers' control. But Boqol Sawm invited us all to join the Muslim Brotherhood. At first it was hard to see how one itinerant preacher could represent a brotherhood, but it was not long before others joined him in the streets around us. And then, with amazing speed, a new mosque was built and Boqol Sawm was installed as its imam. He went from knocking on doors to being the local leader of a movement.

The Muslim Brotherhood seemed like Islam in action. They plucked teenage troublemakers off the streets, put them in madrassas, taught them to pray five times a day, changed their clothes; in fact, changed almost everything about them. I saw just this transformation in the case of the son of one of my relatives. Looking back, I see now that many people embraced the Brotherhood in the first instance simply because they brought order. They did what everyone else believed could not be done: they found a path for these directionless boys who were growing into directionless men. But how exactly did they achieve this feat?

The overarching message of Boqol Sawm was that this life is temporary. If you lived outside the dictates of the Prophet you would burn in hell for the duration of your real life, the afterlife. But if you lived righteously, Allah would reward you in paradise. And men in particular would receive special blessings if they became warriors for Allah.

This was not the practice of my mother, much less my father. No longer were we merely people put on earth to be tested, fearing judgment and asking God to be patient with us. Now we had a task and a goal: we were bound together in an army; we were soldiers of God, fulfilling his purpose. Together, in their different ways, Sister Aziza and Boqol Sawm were the vanguard of a militant Islam—a version that emphasized the political ideology of Muhammad's Medina years (indeed, Boqol Sawm had been trained in Medina). And I fervently embraced it.

Thus, when Ayatollah Khomeini in Iran called for the writer Salman Rushdie to die after he published *The Satanic Verses*, I didn't ask if this was right or what it had to do with me as an expatriate Somali in Kenya. I simply agreed. Everyone in my community believed that Rushdie had to die; after all, he had insulted the Prophet. My friends said it, my religious teachers said it, the Qur'an said it, and I said it and believed it, too. I never questioned the justice of the fatwa against Rushdie; I thought it was completely moral for Khomeini to ensure that this apostate who had insulted the Prophet would be punished, and the appropriate punishment for his crime was death.

The Islam of my childhood, though all encompassing, had not been overtly political. During my teenage years, however, fealty to Islam became something that went far beyond the observance of daily rituals. Islamic scripture, interpreted literally, was presented as the answer to all problems, political, secular, and spiritual, and my friends as well as my own family began to accept this. In the mosques, the streets, and behind the walls of our homes, I saw the established leaders who emphasized the importance of ritual observance, of prayer, fasting, and pilgrimage—the kind of people I have called Mecca Muslims—being replaced by a new breed of charismatic and fiery imams, inspired by Muhammad's time in Medina, who urged action, even violence, against the opponents of Islam: the Jews, the "infidels," even fellow Muslims who neglected their duties or violated the strict rules of sharia. Thus I witnessed the rise of a political ideology wrapped in a religion.

The Medina Muslims are neither spiritual nor religious in the Western sense. They see the Islamic faith as transnational and universal. They prescribe a set of social, economic, and legal practices that are very different from the more general social and moral teachings (such as calls to practice charity or strive for justice) that are found not just in Islam but also in Christianity, Judaism, and other world religions.

Even this might not be so bad if the Medina Muslims were prepared to tolerate other worldviews. But they are not. Their idea is of a world in service to Allah and governed by sharia as exemplified in the *sunnah* (the life, words, and deeds of the Prophet). Other faiths, even other interpretations of Islam, are simply not valid.

My Apostasy

My long and winding journey away from Islam began with my own childish propensity to ask questions. In many respects, I was always a kind of "protestant"—in the sense that I began by protesting against the subordinate role that I, as a girl, was expected to accept. At the age of five or six, I remember asking: "Why am I treated so differently from my brother?" That question prompted the next one: "Why am I not a boy?"

As I grew older, I questioned more of what I heard. Had anyone ever been to hell? Could anyone tell me that it was in fact a real place, which appeared to those condemned to it exactly as it was described in the Qur'an?

"Stupid girl, stop asking so many questions!" I can still hear those words from my mother, my grandmother, my Qur'an teachers, sometimes followed by a slap with the back of a hand. Only my father tolerated inquiry. For her part, my mother simply became convinced that I was bewitched. To doubt, to question, made me in her eyes "feeble in faith." The exercise of my reason itself was forbidden. But the questions never stopped coming, eventually leading to this one: "Why would a benevolent God set up the world like this, marking one half of the population to be second-class citizens? Or was it just men who did this?"

But even those questions were just the first hesitant steps down a long road. My next and perhaps biggest step away from Islam came after an answer—one my father gave—rather than a question.

In January 1992, my father raced to my mother's flat after Friday prayers at the mosque. A man had offered to marry one of his daughters, and he had named me. The man, Osman Moussa, was a member of our clan who was living in Canada. He had returned to Nairobi to choose a bride from among his extended family members. He had his pick of the Westernized Somali girls living in Canada, but he wanted a traditional girl. And with a civil war then raging in our country, Somali brides in Nairobi could be had for free. I was traded in less than ten minutes; Osman Moussa would establish a bond with the Magan family, my father's lineage, and my father would now be able to claim relatives in Canada. It was a simple transaction, part of that system of kinship relations that has governed Somalia—and much of the rest of the world—since time immemorial.

When we were introduced, my intended husband told me that he wanted six sons. He spoke in half-learned Somali and half-learned English. I told my father I did not want to marry him; he replied that the date had been set. What I did not have to do was consummate the marriage. That would wait until I had journeyed to Canada. The air ticket was eventually bought, too. I would be going by way of Germany.

I did not leave Kenya until July. When I arrived in Germany, I walked around the clean streets of Düsseldorf, pondered my options carefully, and shortly thereafter took a train from Bonn to Amsterdam, claiming to be a Somali asylum seeker fleeing the civil war, but in reality fleeing my arranged marriage and the wrath of my family and clan for breaking the marital contract my father had made.

I have told my story at length in my memoir *Infidel*, so here I can be brief. I ended up at a refugee screening camp, was granted asylum, worked hard to get off welfare and learn Dutch, received a university degree, and ended up writing, debating, and then being elected to the Dutch Parliament. What is relevant here is my gradual exit from Islam.

When I arrived in Holland in 1992 I was still a believing and practicing Muslim. I began to shed the practicing part of my faith first. Even so, I was constantly bargaining with myself, finding ways to create circular proofs that I was still a believing, obedient, devout Muslim. When I sent photos home to my family, I made sure to dress with the utmost modesty and to cover my hair. In January 1998, when I rushed back to

Nairobi because of my sister's death, I dug up my old clothes and when I knocked on my mother's door I was dressed pretty much like all the other Somali women there. With my mother and my brother I prayed the required five times a day for the duration of my weeklong visit. As soon as I returned to Holland I switched back to my nonobservant state.

I didn't recognize this distancing immediately; it was only clear in hindsight. If you had asked me anytime between 1992 and 2001, I would have told you I was living as a Muslim. Yet even though I still thought of myself as a Muslim, I developed a lifestyle not much different from that of an ordinary Dutch woman in her twenties. I prioritized study and work over worship; when I made future plans I dropped the *inshallah* (God willing) from my speech. In my free time I pursued fun and recreation.

In addition to neglecting prayer, fasting, and the prescribed Muslim attire for women (the hijab), I proceeded to violate at least two of the six major Qur'anic *hudood* restrictions. The *hudood* prescribe fixed punishments for the consumption of alcohol, illegal sexual intercourse (fornication and adultery), apostasy, theft, highway robbery, and falsely accusing someone of illicit sexual relations. For five years I lived together with my boyfriend, an infidel, out of wedlock, and even talked of having children under that arrangement. And I consumed wine seemingly with the same nonchalance as my Dutch friends.

In reality, though, I was leading a double life. I suffered frequent bouts of guilt and self-condemnation, feeling sure that I was doomed. These feelings were always set off by contact with fellow Muslims—in particular, individuals who took it upon themselves vocally to "command right and forbid wrong," one of the central tenets of Islam (about which more later). My solution was to avoid such people as much as possible, even the Muslims who quietly disapproved. Avoidance was my main strategy to deal with the terrible dissonance between the faith that I purported to believe in and the way I actually lived. It was not easy, but I got better at evasion and, in the years before 9/11, I achieved a kind of peace of mind.

In the months following 9/11, however, it became impossible for me to maintain that fragile balance. I could not overlook the central role the terrorists had attached to the Prophet Muhammad as their source of inspiration, and I was soon openly participating in the debate over Islam's role in the terror acts. When Dutch interviewers directly asked me on live radio and television if I was a Muslim, I minced my words of reply.

Finally, after much agonizing, I resolved my inner conflict by rejecting the claim that God is the author of the Qur'an; by rejecting Muhammad as a moral guide; and by accepting the view that there is no life after death and that God is created by mankind and not the other way around. In doing so I violated the most serious of all the *hudood* restrictions. But there seemed no other option open to me. If I could not submit to Islam, I had to become an apostate.

Yet it would be misleading to suggest that it was 9/11 that led me to question my faith as a Muslim. That was just the catalyst. The more profound cause of my crisis of faith was my exposure prior to 2001 to the foundation of Western thought that values and cultivates critical thinking.

When I was admitted to the University of Leiden, I expected to be presented with a single narrative of events and their significance and one explanation for why everything had happened as it did. Instead, the professors began every course with a central question; spent a lot of time on definitions and their importance; then presented key thinkers and their critics over time. My job as a student was to grasp the central question; to learn about the thinkers, their theories of power, political elites, mass psychology and sociology, and public policy; the methods by which they got to their conclusions; their critics and their methods of criticism. The point of all these exercises was to learn to improve on old ways of doing things through critical thinking. We were graded not just on our factual knowledge, but on our ability to scrutinize any given idea. In this context religion was just another idea, another belief system, another hypothesis, another theory. A critical approach to the words of Jesus was to be no different from a critical approach to the words of Plato or Karl Marx.

My next course, Western Political Thought, included a discussion of the Catholic Church, the Reformation, and the Counter-Reformation. We examined the debate over man-made versus God-made laws. I remember listening, half fascinated and half terrified, because at the time I didn't even want to entertain the idea that man-made laws could supersede God's. I sometimes justified this fascination by saying, well, if it wasn't God's intention for me to be at Leiden then I wouldn't be at Leiden, so I might as well read on.

The more I considered the world around me, the more I began to take issue with all that I had been taught in my previous life. In the Netherlands, for example, I was stunned by the near-total absence of violence. I never saw Dutch people engaging in physical confrontations. There were no threats or fear. If two or three people were killed, it was considered a crisis of the social order and spoken about as such. Two or three violent deaths in my Somali homeland were considered completely ordinary and unremarkable.

Along with the absence of violence, I was overwhelmed by the level of human generosity. Everybody in the Netherlands had medical insurance. In the early 1990s, when I first came to Holland, the Dutch centers where asylum seekers were received were like resorts, with tennis courts, swimming pools, volleyball courts. All our needs—food, medicine, shelter, warmth—were attended to. And on top of that we were also offered psychological assistance and support as part of the universal health-care package that covered every Dutch citizen. The Dutch, I saw to my amazement, took care of everyone who ended up inside their borders, including people who had no connection to Holland, other than hoping it would be a place of refuge.

But the thing that stunned me the most was Holland's approach to gender relations. There were women on television, and they did not wear headscarves, but instead donned fashionable clothes and makeup. Parents raised their girls in exactly the same way as their boys, and girls and boys mingled in school and on the streets. It was the kind of gender mixing that, in the culture I came from, was deemed to be a catastrophe and a sure sign of the approach of the end of days. Here it was so routine that the Dutch people I knew were surprised at my surprise.

Life in the West wasn't perfect, of course. I saw people who were unhappy: white, wealthy people who were disgruntled with their lives, with their work, with their friends, and with their families. But I wasn't much interested in abstract notions like happiness back then; I was simply fascinated by how it was even possible to achieve this level of political stability and economic prosperity.

After 9/11, I began to reexamine the world I had grown up in. I began to reflect that all over this world—in Somalia,

Saudi Arabia, Ethiopia, Kenya, and even inside the Muslim immigrant community in Holland—Islam represented a barrier to progress, especially (but not only) for women. Besides, expressing my doubts about Islam meant that I had no spiritual home: in Islam you are either a believer or a disbeliever. There is no cognitive room to be an agnostic. My family and some of my Muslim friends and acquaintances gave me that stark choice: you are either one of us, in which case you quit voicing your thoughts on Islam, or you are one of the infidels and you get out. And ultimately that was why I could not stay in the religion of my father, my mother, my brother, my sister, and my grandmother.

I was not surprised at all that the Medina Muslims condemned me and wanted me to suffer the "appropriate" punishment for leaving the faith: namely, death. Twelve years before, after all, I had wanted no less for Salman Rushdie. What was far more confusing and grating was the outright hostility of individuals who, just like me prior to my apostasy, had routinely violated other central tenets of the *hudood* in their personal behavior, but who now saw fit to brand me as a traitor to their faith because I no longer wanted to be a sham Muslim. Many secular non-Muslim intellectuals were also quick to dismiss me as a "traumatized" woman working out my own personal demons. (Some continue to make that patronizing claim, like the eminent American journalist who once speculated that my family was "dysfunctional simply because its members never learned to bite their tongues and just say to one another: 'I love you.'")

I was stunned and disheartened to discover that, in this particular debate, one of the core principles of Western liberal achievements—critical thinking about all belief systems—was not to be applied to the faith I had grown up in.

Why I Am Not Exceptional

For years I have been told, condescendingly, that my critique of Islam is a consequence of my own uniquely troubled upbringing. This is rubbish. There are millions of impressionable young men and women like me who have succumbed to the call of the Medina Muslims as I did when I was sixteen. And I believe there are just as many who now yearn to challenge the ultimately intolerable demands that ideology makes on them. In this chapter, I have briefly recounted the story of my early life not because it is exceptional but because I believe it is typical.

Take the case of Shiraz Maher, an idealistic young man who was studying in Leeds, England, on September 11, 2001. Maher had spent his first fourteen years in Saudi Arabia, where the act of wearing a Daffy Duck T-shirt with the words "I Support Operation Desert Storm" (to remove Saddam Hussein from Kuwait) earned him a lecture on the American plot to establish military bases on "holy soil." In 2001, having learned his lesson, he joined Hizbut Tahrir—Arabic for "The Party of Liberation"—which advocates the creation of a caliphate, and duly rose to become one of its regional directors. Maher later described Hizbut Tahrir's philosophy: "It applauds suicide bombers but believes suicide bombing is not a long-term solution."[1]

Where had he learned this philosophy? The answer is that in 1994 Maher had attended a Hizbut Tahrir conference in London where Islamists from Sudan to Pakistan came to talk about forming a caliphate. At the time, no one in the West objected, if indeed they noticed, and certainly no one within the immigrant Muslim community resisted. The result, according to Maher, was that soon the "idea of having an Islamic state had been normalized within the Muslim discourse."[2] This message was spread by a new wave of preachers, who laid an uncompromising emphasis on Muhammad's message in Medina about what Islam was and how it should be practiced. As within my own Somali community in Nairobi, young Muslims in the West were quite easily seduced by the Medina Muslims and their violent call to arms.

Just as I left Islam after 9/11, Maher left Hizbut Tahrir after the 2005 bombing of the London Underground. (He didn't personally know the subway bombers but, like him, they came from Leeds.) My mind had been opened at Leiden; Maher, by contrast, says he had encountered a more pluralistic view of Islam as a graduate student at Cambridge University. Today he is a senior fellow at the International Center for the Study of Radicalization, King's College London, researching the lives of young jihadists.

The problem is that, right now, too many young Muslims are at risk of being seduced by the preaching of the Medina Muslims. The Mecca Muslims may be more numerous but they are too passive, indolent, and—crucially—lacking in the intellectual vigor needed to stand up to the Medina Muslims. When individuals are lured away from their midst by preachers calling for jihad and those individuals then commit an atrocity crying "Allahu Akbar" (God is great), the Mecca Muslims freeze in denial, declaring that the atrocity is un-Islamic. This attempt to decouple the principle from its logical outcome is now something of a joke not only among non-Muslims who mock the "religion of peace" (or the "religion of pieces") but also among Medina Muslims, who openly express contempt for those Muslim clerics who declare that the "peaceful" Meccan verses of the Qur'an somehow abrogate the later and more violent Medina ones.

Consider Tamerlan and Dzhokhar Tsarnaev, the accused Boston Marathon bombers. Growing up, the brothers were typical of Mecca Muslims: they rarely observed Islamic strictures: one had dreams of becoming a boxing champion and spent most of his days training while the other had a busy social life, dated girls, and smoked pot. The parents—at least in their early years in the United States—do not seem to have been very devout. When Dzhokhar, a graduate of the prestigious Rindge and Latin School in Cambridge, Massachusetts, wrote a bloodstained note in the final hours before his capture, the first words he invoked were the same words that I first learned from my grandmother as a very small child: "I believe there is no God but Allah and Muhammad is His messenger."[3] As we have seen, that is the Shahada, the Muslim profession of faith, and it is the most important of the five pillars of Islam. Today the Shahada is the banner of IS, Al-Qaeda, and Boko Haram. It is also the banner of Saudi Arabia, the country that has used so much of its wealth to spread to every corner of the world the practice of Islam in Medina fourteen centuries ago.

Embracing violent jihad has become an all-too-common means for young Muslims to resolve the cognitive pressures of trying to lead an "authentic" Muslim life within a permissive and pluralistic Western society. As we saw earlier, many first-generation Muslim immigrants to the West opt to cocoon themselves and their families, trying to put a wall between themselves and the society around them. But for their children this is simply unsustainable. For them, the choice becomes a stark one between abandoning their faith or embracing the militant message of Medina. "If I were younger and instead of 9/11 it was the Syrian conflict," Maher recently admitted, "there's a very, very good chance I would go. Instead of studying them, I would be the one being studied."[4]

These pressures are not going away. The question is whether or not a third way exists. Must all who question Islam end up either leaving the faith, as I did, or embracing violent jihad?

I believe there is a third option. But it begins with the recognition that Islamic extremism is rooted in Islam itself. Understanding why that is so is the key to finding a third way: a way that allows for some other option between apostasy and atrocity.

I left Islam, and I still think it is the best choice for Muslims who feel trapped between their conscience and the commands of Muhammad. However, it is unrealistic to expect a mass exodus from Islam. This fact leads me to think of the possibility of a third option. A choice that might have enabled someone like me to remain a believer in the God of my family. A choice that might somehow have reconciled religious faith with the key imperatives of modernity: freedom of conscience, tolerance of difference, equality of the sexes, and an investment in life before death.

But in order for that choice to become possible, Muslims have to do what they have been reluctant to do from the very beginning—and that is to engage in a critical appraisal of the core creed of Islam. The next question that has to be addressed is why that has proved so incredibly difficult. After all, I am far from being the first person to call for a Reformation of the religion of my birth. Why have all previous attempts at a Muslim Reformation come to nothing? The answer lies in a fundamental conflict within Islam itself.

CHAPTER 2

WHY HAS THERE BEEN NO MUSLIM REFORMATION?

In 2012, the Harvard Kennedy School invited me to lead a study group dealing with the intersection of religion, politics, society, and statecraft within the Islamic world. I have now done this for three years. Its focus is on Islamic political theory. The seminar is geared toward mid-career students ranging in age from their mid-twenties to their forties, but undergraduates can also participate. Our meetings last for ninety minutes, and there is an ample reading list.

As must now be clear, I have been an uncompromising critic of political Islam for more than a decade. But in recent years I had come to feel that rather than simply inveighing against it, I must reengage with Islam—the religion as well as the ideology—not only to deepen my understanding of its complex religious and cultural legacy, but also for the sake of those who find themselves, as I once did, trapped between the demands of a rigid faith and the attractions of a modern society. This book is one of the fruits of that decision. As such it represents a continuation of the personal and intellectual journey I have chronicled in my previous books. The study group was a crucial preliminary step.

From the start, I was curious about the students who signed up for my study group. The first class list that I received from the registrar's office provided a range of names, some clearly Anglophone, some clearly Arabic. About half were Americans, two of whom were members of the U.S. military, and most of whom had worked or served in Islamic countries. At least three of the Americans were Jewish. The rest of the class were nearly all Muslim: men from Qatar, Turkey, Lebanon, Pakistan, and Senegal, along with a young woman from Niger. In many ways, the class's Muslim students were a microcosm of the modern Muslim elite: they were educated, mobile, frequently wealthy, and held a variety of views on Islam. However, it quickly became clear that some of the attendees thought that there could be no other view than their own.

On that first afternoon, the students assembled, we made our introductions, and I began to speak. I got as far as the first few sentences when the Qatari student raised his hand and began addressing the rest of the room. He said that he needed to "clarify" what I was saying. Then another—the Pakistani—interrupted. A third and then a fourth chimed in. For any remark that I made involving Islam, one of them had a clarification. And, almost from the first word, they got personal. According to one of them, I was a "traumatized woman projecting my personal experience and brainwashing people." Another wanted everyone to understand that I was just "an Islamophobe telling lies."

Most of the other students (including the other Muslim students) were stunned. It was, for a while, a bit of a tennis match—heads swiveled, following their verbal volleys and my efforts to return. But, as the minutes passed, the tension within the class grew greater. It was not necessarily that the other students did not want to speak; it was that they could not get a word in. And it was not just the first session that went like this. It was the same week after week—until the fourth week, when the malcontents ceased to attend.

I have no problem with discussion and debate. That was the point of the course. Yet these days it is too short a journey from preemptively challenging any critic of Islam, to correcting them, implying threats, and silencing them outright. To my mind, nothing could be more "clarifying" of the fundamental problem facing Islam today than those painful early sessions in the seminar room.

I had not designed the course to be a seminar on my personal vision of Islam. I had been careful not to assign my own writings. Instead, I had drawn up a balanced list of scholarly articles and academic books, points and counterpoints around the nature of political theory in Islam. This material was what I had intended to discuss in class. Yet it was as if the objectionable students had not even looked at the syllabus. For them, simply to ask a question about Islam was a grave offense.

So, to start with, we need simply to ask: Why is it so hard to question anything about Islam? The obvious answer is that there is now an internationally organized "honor brigade" that exists to prevent such questioning. The deeper historical answer may lie in the fear of many Muslim clerics that allowing critical thought might lead many to leave Islam. Yusuf Al-Qaradawi, a staunch Medina Muslim and a prominent leader of the Muslim Brotherhood, has said: "If they had gotten rid of the apostasy punishment Islam would not exist today. Islam would have ended with the death of the Prophet, peace be upon him. Opposing apostasy is what kept Islam to this day."[1] The clerics fear that even the smallest of questions will lead to doubt, doubt will lead to more questions, and ultimately the questioning mind will demand not only answers but also innovations. An innovation in turn will create a precedent. Other minds that question will build on these precedents and more concessions will be demanded. Soon people will be innovating themselves out of their faith altogether.

Innovation of faith is one of the gravest sins in Islam, on a par with murder and apostasy. Thus it is perfectly intelligible why the leading Muslim clerics (the ulema) have come to the consensus that Islam is more than a mere religion, but rather the one and only comprehensive system that embraces, explains, integrates, and dictates all aspects of human life: personal, cultural, political, as well as religious. In short, Islam handles everything. Any cleric who advocates the separation of mosque and state is instantly anathematized. He is declared a heretic and his work is removed from the bookshelves. This is what makes Islam fundamentally different from other twenty-first-century monotheistic religions.

It is important to grasp the extent to which religion is intertwined with politics and political systems in Islamic societies. It is not simply that the boundaries between religion and politics are porous. There scarcely are any boundaries. Seventeen Muslim-majority nations declare Islam the state religion and require the head of state to be a practicing Muslim, while in the Christian world only two nations require a Christian head of state (although the British monarch is required to be the "Defender of the Faith," the heir to the throne intends to be "Defender of Faith").[2] In countries like Saudi Arabia and

Iran, or within mounting insurgent movements such as IS and Boko Haram, the boundaries between religion and politics do not exist at all.

This fusion of the spiritual and the temporal offers an initial clue as to why a Muslim Reformation has yet to happen. For it was in large measure the separateness of church and state in early modern Europe that made the Christian Reformation viable.

The Lesson of Luther

Will a Muslim Reformation look exactly like the Christian one? No, of course not. But there are some important resemblances, and it is these that give me hope.

In October 1517, a somewhat obscure but very obstinate monk in the Saxon town of Wittenberg wrote ninety-five theses decrying the Church's practice of selling indulgences for salvation. His name was Martin Luther and his words helped trigger both a theological and a political revolution.

The history of the Protestant Reformation is complex and must be heavily simplified here. Three crucial points stand out. First, unlike previous European heretics, Luther was able to exploit a new and powerful technology to spread his message: the printing press. Second, his key ideas—such as "justification by faith alone" and "the priesthood of all believers"—appealed strongly to a new and growing class of city-dwellers, whose literacy and prosperity made them impatient of the corrupt practices of the Roman Church. Third, and crucially, it was in the interests of a significant number of European states—among them England—to back Luther's challenge to the pope's ecclesiastical hierarchy.

The upshot was a huge upheaval. Not only was Western Christendom irrevocably split between Protestants and Catholics. After more than a century of bloody religious wars within and between states, a new order was established that gave primacy to secular authority over religious (the principle of *cuius regio, eius religio* essentially left it to each of the various European princes to choose the faith of his realm).[3] Yet after the dust settled, the Western world was utterly transformed, with the Protestant nations often leading the way in the invention of new social, political, and cultural forms.

The German sociologist Max Weber argued in his landmark work, *The Protestant Ethic and the Spirit of Capitalism*, that Reformation theology encouraged the godly to look for signs of divine grace in the success of their worldly pursuits. The sanctification of thrift and the cultivation of "capitalistic" virtues, he argued, fueled an economic revolution. Perhaps; or it may simply be that the universal literacy promoted by Protestantism spurred learning and productivity. Either way, from the middle of the seventeenth century, the Western world began an astonishing sequence of intellectual as well as economic and social revolutions: the Scientific Revolution, the Enlightenment, the Industrial Revolution, the American and French Revolutions. From here, we can trace not only the rise of modern science, but also the rise of capitalism and representative government, with its ideals of self-governance, tolerance, freedom, and equality before the law. Out of the changes wrought by the Reformation—in particular its emphasis on universal literacy—came a remarkable number of the things that made us modern.

In short, the liberation of the individual conscience from hierarchical and priestly authority opened up space for critical thinking in every field of human activity.

Centuries later, Islam has had no comparable awakening. The golden age of Islamic science and philosophy, which predated the European Enlightenment, lies a thousand years in the past. While many Muslim nations have benefited from advances in science and economics, while they now have their gleaming skyscrapers and infrastructure, the *philosophical* revolution that grew out of the Protestant Reformation has largely passed them by. Instead, much of the Muslim world, both inside Muslim-majority nations and in the West, lives half in and half out of modernity. Islam is content to use the West's technological products—there is even an app that will remind you when to say your five daily prayers—but resists the underlying values that produced them. (This, of course, helps explain the notorious lack of scientific and technological innovation that characterizes the entire Muslim world.)

This is not to say that there have not been sporadic attempts at change. As long ago as the eighth century, there were repeated efforts within Islam to incorporate ideas from Greek philosophy to make the religion less all-encompassing and inflexible in its demands upon believers. In the eighth to tenth centuries, for example, the Mu'tazila school of Islamic thought, which proclaimed the dignity of reason and argued that Islamic doctrine should be open to contemporary interpretation, flourished in Baghdad. But it was resoundingly defeated by the Ash'ari school, led by Imam Ash'ari, a former Mu'tazila believer who argued with the usual zeal of a convert that the Qur'an was the perfect and immutable word of God. The triumph of the Asha'ri school cemented a belief that, with the message of Muhammad, "History came to an end." And that has been the endpoint for most debates within Islam down to our own era. Indeed, something very similar happened in the twentieth century.

We are constantly reminded that, at the start of the twentieth century, the Islamic, and particularly the Arab, world had a wide range of independent political publications and literary and scientific journals through which it was possible to exchange ideas and import advances from the West. The mid-nineteenth-century Syrian political thinker Francis Marrash, who hailed from Aleppo and studied medicine in Paris, had published writings about the importance of freedom and equality and the vital role to be played in the modernization of Arab society by education and "a love of country free from religious considerations."[4] This was not completely delusional. By the end of World War II, the central features of sharia had been replaced in many Muslim countries by laws based on European models. Polygamy was legally abolished, civil marriage introduced. Arabs were also embracing nationalism as well as a belief in the importance of pre-Islamic Arabic culture.

At the same time, Islam itself was increasingly being reinterpreted as part of a long continuum in man's attempts to achieve social justice, even being used at times to validate socialist doctrines of redistribution and other efforts to remake society. An Egyptian thinker named Khalid Muhammad Khalid declared that true religion was possible only when social and economic justice existed, and he proposed among other things nationalizing natural resources, dividing up large estates, instituting labor rights, and fixing agricultural rents, as well as emancipating women and providing birth control. Other early-twentieth-century Muslim thinkers sought to reassess the linkages between seventh-century Islamic law and the modern state. In the twentieth century, men such as Ali Abdel Raziq, Mahmoud Mohammed Taha, Nasr Abu Zayd, and Abdolkarim Soroush—all Islamic thinkers—proposed fundamental reforms.

Though few people today know the names of these men, their proposals and the ensuing responses have much to teach us.

Ali Abdel Raziq, an Oxford-educated Egyptian scholar and a professor at Al-Azhar University, was a devout Muslim and religious judge who argued that Islam should be completely separated from politics so as to protect it from political corruption. In his 1925 book, *Islam and the Foundations of Governance*, Abdel Raziq argued that Muslims could use their innate powers of reason to devise the political and civil laws best suited for their times and circumstances. What is more, he specifically rejected the idea of restoring a Muslim caliphate, so dear to modern radicals. "In truth," he wrote:

> This institution which Muslims generally know as the caliphate has nothing to do with religion. It has . . . more to do with . . . the lust for power and the exercise of intimidation that has been associated with this institution. The caliphate is not among the tenets of the faith. . . . There is not a single principle of the faith that forbids Muslims to co-operate with other nations in the total enterprise of the social and political sciences. There is no principle that prevents them from dismantling this obsolete system, a system which has demeaned and subjugated them, crushing them in its iron grip. Nothing stops them from building their state and their system of government on the basis of past constructions of human reason, of systems whose sturdiness has stood the test of time, which the experience of nations has shown to be effective.

For positing these ideas, Abdel Raziq was dismissed from Al-Azhar. The university's Supreme Council condemned and denounced his book, and expelled him from the circle of the ulema. He lost his title of *alim*, or learned man, and was forced into domestic exile, escaping a worse fate thanks only to his family's prominence.

Three years later, a new group began to emerge in Egypt under the leadership of a schoolteacher named Hassan al-Banna. Disgusted by what he believed was an excess of materialism and secularism, as well as the sight of Egyptians laboring for foreign bosses, al-Banna wanted a return to a precolonial era, when religion had been a comprehensive way of life—although he himself was largely self-taught and did not come from a learned, clerical background. Instead of fostering a new secular nationalism consistent with developments in Europe and elsewhere in the modern world, al-Banna wanted Muslims everywhere to join together in a larger community founded upon Islam and Islamic religious law. In al-Banna's vision of the Islamic state, there would be no political parties, sharia would form the legal code, and only those who had a religious education would rule or administer the government. Schools themselves should be attached to mosques. In this way, Islam would be the guiding, unifying principle across the Arab Muslim world.

Hassan al-Banna is hardly a household name in the West, but the organization that he helped to found has become one: the Muslim Brotherhood. And his writings inspired some of the most familiar names of the late twentieth and early twenty-first century, among them Ayatollah Khomeini and Osama bin Laden.[5]

The triumph of al-Banna over Abdel Raziq—in essence, the triumph of theocracy over reform—can also be seen in the fates of other twentieth-century Islamic reformers. The Sudanese intellectual Mahmoud Mohammed Taha argued that Muslims should embrace the spiritual Islam of Mecca and let go of the Islam of Muhammad's more warlike and political Medina period, which, Taha argued, applied only to that specific moment in time and not to subsequent generations. Taha also campaigned against introducing sharia in Sudan. Though he still believed there was no god but Allah, and that Muhammad was his messenger, Taha was nonetheless hanged for apostasy in 1985.

More recently, Nasr Abu Zayd, an Egyptian thinker, argued that human language had at least some role in shaping the Qur'an, thus making it not completely the uncorrupted word of Allah. For proposing a reinterpretation of the sacred text, he was deemed an apostate by an Egyptian court in 1995, and then forcibly divorced from his wife against his (and his wife's) will, because he was now a non-Muslim, and a non-Muslim man cannot be married to a Muslim woman. After receiving death threats, Abu Zayd fled Egypt and went into exile in the Netherlands.

In Iran, the Islamic thinker Abdolkarim Soroush, though he supported the Islamic revolution of 1979, later argued that political power should be far more separate from religious leadership than it is today. For making this argument, Soroush received numerous threats, was forced to end his university teaching, and eventually found life so intolerable that he, too, moved abroad.

All of these would-be reformers based their arguments on Islamic theological grounds. But the ulema have not only resisted all such attempts at reform; they have time and again successfully threatened and bullied the reformers into silence or exile, where they have not actually secured their execution. And the method has been to return, always, to the Qur'an. Because the Qur'an is inviolate, timeless, and perfect, they argue, what is written in it cannot be criticized, much less changed.

That explains why, in Islam, reform has never had positive connotations and innovation is at all costs to be avoided. As Albert Hourani explains, after the appearance of Muhammad, "History could have no more lessons to teach, if there was change it could only be for the worse, and the worse could only be cured not by creating something new but by renewing what had once existed."[6] In other words, "reform" is simply not a legitimate concept in Islamic doctrine. The only accepted and proper goal of a Muslim "reformer" is a return to first principles. The hadith, the text containing the words and deeds of Allah's Prophet, credits Muhammad with saying that his generation would be the best of all, the one that followed him the next best, and so on down.[7] It is the precise opposite of the Western narrative of progress: in this version of history, instead of improving, each generation is worse than the one before. Only when, at the turn of every century, a renovator arrived, a *mujaddid*, could Islam revert back to its moment of perfection at the time of its founding, the time of Muhammad.[8] In those terms, it is only the Medina Muslims who can represent themselves as the agents of a Muslim Reformation.

Today, the most notorious exponent of this kind of "reform," in the sense of restoration, is the self-styled Islamic State in Iraq and Syria, which proposes to create a new caliphate where the only law is sharia. Adulterers there are stoned to death, infidels beheaded, and thieves mutilated. Indeed, much Islamic State propaganda is like a YouTube upload of a time-travel trip back to the seventh century. If these are the people who claim to be purifying Islam, what chance does real reform stand?

Who Speaks for Islam?

Luther's Reformation was launched against a hierarchical ecclesiastical establishment. When the pope sought to anathematize him, Luther could retort: "I am called a heretic by those whose purses will suffer from my truths." Islam is different. Unlike Catholicism, Islam is almost entirely decentralized. There is no pope, no College of Cardinals, nothing like the Southern Baptist Convention—no hierarchical structure, no centrally controlled system of ordination. Any man can become an imam; all it takes is a self-professed knowledge of the Qur'an and followers.

I am always intrigued when on college campuses there are heated demands that an imam or scholar of Islam be present when I speak to offer the "correct" interpretation of Islam. That was the demand of Yale's Muslim Student Association in September 2014, when I was invited to the university's campus to give the Buckley Lecture. But whom did they have in mind for this role? A Saudi cleric? An American convert? An Indonesian? An Egyptian? A Sunni? A Shiite? A representative of Islamic State, perhaps? Or how about Zeba Khan, an American Muslim of Indian descent, who was educated at a Jewish day school while also attending a mosque in Toledo, Ohio, where men and women prayed side by side, and who in 2008 started the group Muslims for Obama? Or perhaps they would prefer the British-born lawyer turned imam, Anjem Choudary, who favors the imposition of sharia in Britain and has looked forward to seeing the black flag of IS flying over Parliament? All can legitimately claim to speak for Islam. There is no Muslim pope to say which of them is right.

In my own Harvard seminar room, a Muslim woman

from Egypt became very argumentative. She came to some sessions of my study group and not to others, but was always ready to contradict whatever I was saying. Finally, I asked her about a point that had been made in the assigned reading. She replied: "I haven't done the assigned reading. I don't need to. I already know everything." This goes to the heart of the matter. Paradoxically, Islam is the most decentralized and yet, at the same time, the most rigid religion in the world. Everyone feels entitled to rule out free discussion.

One of the fiercest critics of my course was a female Sudanese student. Despite never actually attending a single session of the study group, she was completely convinced that everything being said in the classroom was a serious affront to Islam. She was one of a number of Muslim students who lobbied the Kennedy School authorities to have my study group terminated. When one of my colleagues made the point that academic freedom—the freedom to teach and learn about viewpoints and ideas that are fundamentally at odds with others' beliefs—is the cornerstone of the Western university, she reacted with perplexed hostility. Academic freedom was a concept that seemed to her deplorable if it permitted any questioning of her faith.

To understand this hostility, it is important to recognize that the long traditions in Judaism and Christianity of passionate debate and agonizing doubt are largely absent in Islam. There are no great schisms within the Sunni or Shia branches (a division that was not originally theological in nature, but was essentially a dispute over succession). Instead, there is conformity. There is no Reform or Reconstructionist Islam, as there is in Judaism. Rather, like the pre-Reformation Catholic Church, Islam is still persecuting heretics.

Consider this admonition from a Roman Catholic professor of theology, David Bonagura, who notes that Catholic worship is often considered more "stoic" compared with the "energy" of Protestant services, but who goes on to say that these "different styles are pathways to faith," adding that "we need not think our preferred religious experience should be shared by everyone else."[9] How many Muslim clerics today would dare say such a thing?

In no other modern religion is dissent still a crime, punishable by death. When a conservative Jewish rabbi said in a Modern Orthodox Jewish synagogue in Washington, D.C., that Orthodox Judaism needs female rabbis, he was not denounced. A few people in the audience even applauded. When Pope Francis broached the idea of toleration for homosexuals within the Catholic Church, there was heated disagreement, but no violence, and no one called for his overthrow or death.

By contrast, consider the case of Hamza Kashgari, a twenty-three-year-old Saudi man, who in 2013 was accused of blasphemy and threatened with death for having openly challenged the authority of the Prophet Muhammad. What did Kashgari do that was so reprehensible? On the eve of the Prophet's birthday, he addressed a series of tweets directly to Muhammad. In an almost immediate response, Saudi sheiks took to YouTube to demand his execution; a Facebook group demanding his death had ten thousand "friends" within one week—not surprising perhaps when one considers that Saudi Arabia's homegrown Twitter heroes are clerics such as Muhammad al-Arifi, who cannot enter any European nation because of his unabashed support for wife-beating and his hatred of Jews. (Al-Arifi has 10.7 million Twitter followers.)

Kashgari, a newspaper columnist from the port city of Jeddah on the Red Sea, promptly deleted his tweets and fled to Malaysia, where he was detained in the departure hall of Kuala Lumpur International Airport by police as he tried to board a flight to New Zealand. He was soon thereafter repatriated to Saudi Arabia.

What had he written in 140 characters that was so blasphemous? The answer is this:

> On your birthday, I will say that I have loved the rebel in you, that you've always been a source of inspiration to me, and that I do not like the halos of divinity surrounding you. I shall not pray for you.[10]

He also posted: "On your birthday, I find you wherever I turn. I will say that I have loved aspects of you, hated others, and could not understand many more." And finally: "I shall not kiss your hand. Rather, I shall shake it as equals do, and smile at you as you smile at me. I shall speak to you as a friend, no more."[11]

For these innocent words, clerics rose up to demand Kashgari's death for the crime of apostasy, and King Abdullah ordered a warrant for his arrest. It did not matter that Kashgari had apologized and erased his tweets. He was jailed. And although he was freed some eight months later, he has effectively been silenced.

This is a young man who grew up in a conservative religious home, who was doing no more than testing and feeling about the contours of his faith. He did not reject Islam, Allah, or the Prophet. His words merely sought to humanize a religious icon. And for this he was jailed.

The Unexpected Reformation

For many years, Western writers have dreamed of a Muslim Reformation. None has come. Accordingly, most observers of the Islamic world today have given up on the idea. But I believe that a Reformation is not merely imminent; it is now under way. The Protestant Reformation itself erupted quite suddenly. With Islam, with equal suddenness, the change has already begun and will only accelerate in the years that lie ahead.

Recall the three factors that were crucial to the success of the Protestant Reformation: technological change, urbanization, and the interests of a significant number of European states in backing Luther's challenge to the status quo. All three are present in the Muslim world today.

Modern information technology, like the printing press in Luther's time, can certainly be used to promote intolerance, violence, and millenarian visions. But it can also act as a channel for the very opposite things, just as the printing presses of seventeenth-century Europe went from publishing tracts about witchcraft to treatises about physics. The case of Hamza Kashgari in fact perfectly illustrates the way the Internet has the opportunity to be to the Muslim Reformation what the printing press was to the Protestant Christian one. Raised a religious conservative, Kashgari is said to have become a "humanist" under the influence of what he read online.

There is also a constituency for a true Reformation in the Muslim world, just as there was a constituency receptive to Luther's message in sixteenth-century Germany. Muslim city-dwellers are much more likely to be resistant to the people I have called Medina Muslims than people living in the countryside—not least because in practice the imposition of sharia is highly disruptive of a whole range of mainly urban businesses (among them, tourism).

In 2014, the Pew Research Center surveyed more than 14,000 Muslims in fourteen countries. In only two nations, Senegal and Indonesia, was concern about Islamic extremism felt by fewer than 50 percent of the surveyed population.[12] The numbers in the Middle East and North Africa were astounding: fully 92 percent of Lebanese, 80 percent of Tunisians, 75 percent of Egyptians, and 72 percent of Nigerians—huge majorities of people—said they were worried about Islamic extremism. There is good reason to think that it is city-dwellers who are doing most of the worrying.

Moreover, Islam is now a global religion with what might even be called a global diaspora. As a result of postwar migrations, there are more than 20 million Muslims living in Western Europe and North America. These Muslims are, as we have seen, confronting the daily challenge of existing in the modern secular West while still remaining Muslim. In short, there is a rapidly growing potential audience for ideas about a new direction for Islam.

Finally, just as in sixteenth-century Europe, there is now a political constituency for religious reform in key states of the Muslim world. On New Year's Day 2015, to mark the approaching birthday of the Prophet Muhammad, the president of Egypt, Abdel Fattah el-Sisi, gave an astonishing speech at Al-Azhar University itself, in which he called for nothing less than a "religious revolution":

> Is it possible that 1.6 billion people [Muslims] should want to kill the rest of the world's inhabitants—that is 7 billion—so that they themselves may live? Impossible!

> I am saying these words here at Al-Azhar, before this assembly of scholars and ulema—Allah Almighty be witness to your truth on Judgment Day concerning that which I'm talking about now.
>
> All this that I am telling you, you cannot feel it if you remain trapped within this mindset. You need to step outside of yourselves to be able to observe it and reflect on it from a more enlightened perspective.
>
> I say and repeat again that *we are in need of a religious revolution*. You, imams, are responsible before Allah. The entire world, I say it again, the entire world is waiting for your next move . . . because this umma is being torn, it is being destroyed, it is being lost—and it is being lost by our own hands.[13]

El-Sisi is by no means the only Muslim leader who sees the Muslim Brotherhood and its ilk as posing a fundamental threat to his country's political stability and economic development. Similar encouragement of religious reform is being given by the government of the United Arab Emirates.

It is, of course, conventional to argue that el-Sisi's election as president was a symptom of the failure of the Arab Spring. But that is to misunderstand the process unleashed by the revolutions that began in Tunisia in late 2010. The revolutions there, as well as in Egypt, Libya, and Syria, were directed against corrupt dictators; they were then hijacked by Medina Muslims such as the Muslim Brotherhood, whom the dictators had long held in check. When that became clear to Egyptians—especially city-dwellers—they took to the streets once again to oust the Brotherhood government of Mohamed Morsi.

As a challenge to authority—as a revolution against dictators who had once seemed immovable and all-powerful—the Arab Spring was actually a success. It showed that the mighty could be challenged. When another form of authority—religious authority—sought to exploit the opening, there was a second revolution, at least in Egypt (and civil wars in other countries). Eventually, I believe, refusal to submit to the authority of secular rulers will be followed by a more general refusal to submit to the authority of the imam, the mullah, the ayatollah, the ulema.

The ferment we see in the Muslim world today is not solely due to despotic political systems. It is not solely due to failing economies and the poverty they breed. *Rather, it is due to Islam itself and the incompatibility of certain key facets of the Muslim faith with modernity.* That is why the most important conflict in the world today is between those who will defend to the death those incompatibilities and those who are prepared to challenge them—not to overthrow Islam, but to *reform* it.

The initial work of challenging authority has already begun—tragically exemplified by the note written by the son of the newly elected Iranian president shortly before his suicide in 1992: "I hate your government, your lies, your corruption, your religion, your double acts and your hypocrisy."[14] Yet a Reformation cannot be achieved by suicide notes. Like Luther's Reformation, it needs theses: calls for action.

Five Theses

What does one do with a timeworn but historically valuable house? One approach is simply to knock it over and build a new house in its stead. This is not going to happen with

Islam, or any other established religion. A second approach is to preserve the place exactly as it was when first built, unstable and in danger of total collapse though it is. This is essentially the thing that groups such as the Muslim Brotherhood, Al-Qaeda, and IS agree on: a restoration of the seventh-century original.

The third choice is to keep as much as possible of the historic details, make the outside look a lot like the original, but change the house radically from the inside, equipping it with the latest amenities. That is the kind of Reformation or Modification I favor. Extending the metaphor, another term for what I have in mind might be an Islamic Renovation.

I am no Luther. Nor do I have ninety-five theses to nail upon a door. In fact, I have only five. They refer to the five basic tenets of the Islamic faith that those who preach jihad and destruction use with such lethal success. Amending them will, I know, be exceedingly difficult. But for Islam to coexist with modernity, for Islamic states to coexist with other nations on our ever-shrinking planet, and especially for tens of millions of believing Muslims to flourish in Western societies, these five concepts must be amended. Reason and conscience demand it. These changes, I believe, can be the basis of a true Islamic Reformation, one that progresses to the twenty-first century rather than regresses to the seventh.

Some of these changes may strike readers as too fundamental to Islamic belief to be feasible. But like the partition walls or superfluous stairways that a successful renovation removes, they can in fact be modified without causing the entire structure to collapse. Indeed, I believe these modifications will actually strengthen Islam by making it easier for Muslims to live in harmony with the modern world. It is those hell-bent on restoring it to its original state who are much more likely to lead Islam to destruction. Here again are my five theses, nailed to a virtual door:

1. Ensure that Muhammad and the Qur'an are open to interpretation and criticism.
2. Give priority to this life, not the afterlife.
3. Shackle sharia and end its supremacy over secular law.
4. End the practice of "commanding right, forbidding wrong."
5. Abandon the call to jihad.

In the chapters that follow, I will explore the source of the ideas and doctrines in question and evaluate the prospects for reforming them. For now, we may simply note that they are closely interrelated. The main problem for us is obviously the promotion of jihad. But the appeal of holy war cannot be understood without factoring in the prestige of the Prophet himself as a model for Muslim behavior, the insistence on a literal reading of the Qur'an and the attendant rejection of critical thinking, the primacy of the afterlife in Muslim theology, the power of religious law, and the license bestowed on individual Muslims to enforce its codes and disciplines. These issues overlap to the extent that they are sometimes hard to separate. But all must be addressed.

As readers of my previous books will realize, this represents a new approach. When I wrote my last book, *Nomad*, I believed that Islam was beyond reform, that perhaps the best thing for religious believers in Islam to do was to pick another god. I was certain of it, not unlike the Italian writer and Holocaust survivor, Primo Levi, who wrote in 1987 of his absolute certainty that the Berlin Wall would endure. Two years later, the Wall fell. Seven months after I published *Nomad* came the start of the Arab Spring. I watched four national governments fall—Egypt's twice—and protests or uprisings occur in fourteen other nations, and I thought simply: I was wrong. Ordinary Muslims *are* ready for change.

The path forward will be hard, even bloody. But unlike previous waves of reform that foundered on the monolith of religious and political power, today it is possible to find a fellowship of people who desire a separation of religion from politics in the Muslim world.

I am not a cleric. I have no weekly congregation. I simply lecture, read, write, think, and teach a small seminar at Harvard. Those who might object that I am not a trained theologian or historian of Islam are correct. But it is not my purpose singlehandedly to engage the Muslim world in a theological debate. Rather, it is my purpose to encourage Muslim reformers and dissidents to confront obstacles to reform—and to encourage the rest of us to support them in whatever way we can.

For me there can be no going back. It is too late to return to the faith of my parents and grandparents. But it is not too late for millions of others to reconcile their Islamic faith with the twenty-first century.

Nor is my dream of a Muslim Reformation a matter for Muslims alone. People of all faiths, or of no faith, have a great interest in a changed Islam: a faith that is more respectful of the basic doctrines of human rights, that universally preaches less violence and more tolerance, that promotes less corrupt and less chaotic governments, that allows for more doubt and more dissent, that encourages more education, more freedom, and more equality before a modern system of law.

I see no other way forward for us—at least no other way

that is not strewn with corpses. Islam and modernity must be reconciled. And that can happen only if Islam itself is modernized. Call it a Muslim Renovation if you prefer. But whatever label you choose, take these five amendments as the starting point for an honest debate about Islam. It is a debate that must begin with a reconsideration of the Prophet and his book as infallible sources of guidance for life in this world.

CHAPTER 3

MUHAMMAD AND THE QUR'AN

How Unquestioning Reverence for the Prophet and His Book Obstructs Reform

A key problem for Islam today can be summarized in three simplifying sentences: Christians worship a man made divine. Jews worship a book. And Muslims worship both.

Christians believe in the divinity of Jesus while also stating that the Christian Bible was written by men. Jews believe in the sanctity of the Torah, which they kiss and treat with reverence during their services; but they traditionally ascribe its authorship to Moses, a prophet who, like other Hebrew prophets, is presented as human and fallible. However, Muslims believe in both the superhuman perfection of Muhammad and the literal truth and sanctity of the Qur'an as the direct revelation of God. Indeed, while even Orthodox Jewish rabbis argue that it is impossible to defile the Torah,

Muslims believe the opposite—so much so that the charge of disrespecting Muhammad or the Qur'an is enough to incite violent protests, riots, and, frequently, death.

For example, erroneous charges in 2005 that U.S. guards had flushed a Qur'an down the toilet in the Guantánamo Bay detention center resulted in violent riots in many Muslim nations. Seventeen people died in Afghanistan in the ensuing rage and frenzy. More recently, in November 2014, a Christian man and his wife living in Lahore, Pakistan, were beaten and burned alive in a brick kiln after they were accused of burning pages of the Qur'an. (The couple protested their innocence.) Likewise, a series of twelve satirical cartoons depicting the Prophet, which were published in the Danish newspaper *Jyllands-Posten* in September 2005, triggered a paroxysm of outrage across the Muslim world that resulted in more than two hundred reported deaths as well as attacks on Western embassies.

These episodes reflect a key distinction between the West and the Muslim world. While an irreverent approach to religious figures and beliefs is tolerated and even encouraged in Western societies, Muslims regard any "insult" to the Prophet or the Qur'an as deserving the ultimate penalty. And this is not an extreme position. As I mentioned earlier, as a teenager I myself unthinkingly agreed that Salman Rushdie deserved to die for writing a novel that very few people in the Muslim world, myself included, had read.

To understand the roots of the problem, and why I believe that it is not in fact insoluble, we need to reexamine Islam's two most sacred elements: its Prophet, and its holy book. Muslims need to understand Muhammad as a real man, in the context of his times, and the Qur'an as a historically constructed text, not as a divine instruction manual for life today.

Who Was Muhammad?

He is the greatest lawgiver of all time. The revelations he received, along with the facts of his life, form the foundation of a legal code that governs hundreds of millions of people. Yet scholars cannot agree on which year or on which date he was born. The most commonly accepted time is 570 years after the birth of Jesus Christ. His father died before he arrived in the world; by the age of six he had become an orphan. An uncle raised him. He met his first wife when she hired him to act as her commercial agent on a trading mission to Syria. A servant informed her that two angels had watched over the young agent as he slept, and that he had rested under a tree that was known to offer shade only "to prophets."

The young agent was twenty-five, his employer was forty. It was his first marriage and her third, and she initiated the wedding proposal. It would be another fifteen years before the words that would eventually become the Qur'an were first revealed to him. His wife, Khadija, was his first convert.

Over the next twenty-two years, the man known as Muhammad would establish the world's last great religion, create an intertwined religious, political, and legal order, and plant the seeds of an empire that would stretch from the Asian steppes to northern Africa and up through the Iberian peninsula. Today, more than a billion people profess their faith by saying the Shahada—"There is no God but Allah, and Muhammad is His messenger." In nearly fourteen hundred years, that message has remained unchanged.

What made this message revolutionary was not simply the belief in one God, as opposed to the worship of many. This was hardly original, and indeed Muhammad presented his religion as the extension and fulfillment of the monotheistic revelations of Abraham, Moses, and Jesus. What made Islam revolutionary was its vast scope, extending well beyond theology. Islam, as Muhammad devised it, is not simply a religion or a system of worship. It is, as the social anthropologist Ernest Gellner has put it, "the blueprint of a social order."[1] In its very name, "Islam" means submission. You subsume yourself to an entire system of beliefs. The rules as set down are precise and exacting.

Islam became so multifaceted and all-encompassing in part because Muhammad and Islam were a prophet and a faith for their place and time. Muhammad is usually understood in his familiar roles as warrior and prophet. But it is in some ways more revealing and interesting to view him in another role—that of a tribal leader. Muhammad's achievement in this capacity was to create a new religiously based community out of the loosely organized elements of tribal Arab society. In short, he was as much the founder of a "supertribe" as a religious and military figure.

There is general agreement that Muhammad existed, though little is known for certain of his life. But while we cannot verify the facts of his biography, what can be surmised is that he was a product of the kin-based social order that then prevailed throughout the Middle East.

Before Islam, there was kinship. Families, clans, and tribes are the basis of organization in all pre-state societies. The basic social unit is the lineage, a group of families united by their descent from a common ancestor. Each family is part of a lineage; many lineages make up a clan; many clans make up a tribe. All in turn are thought to be descended from a single (mythological or semidivine) founder.

But while they are united by the fiction of common descent, these kin groups are decentralized and fractious, frequently riven by feuds that can go on for generations. Strong leadership is needed to unite them if they are not to degenerate (as they did in the West) into mere shared names with next to no bonds of mutual allegiance. This was the case in Muhammad's time. It was still the case fourteen hundred years later when T. E. Lawrence united the Bedouin tribes against the Turks in World War I. It was also true of my own native Somali environment.

In this world of shifting interests and allegiances, tribal leaders arise through personal qualities of strength, cunning, and innate magnetism. The tribal leader plays many roles: he is lawgiver and judge, businessman, war chief, and head of the tribe's religious cult. He is also a source of patronage and distributes the bounty of commerce and war. Honor and personal loyalty (often reinforced by strategic marriages) are the primary bonds that support the tribal leader and hold the system together. Based on what we know of him from Islamic sources, Muhammad fulfilled all these roles. He transcended tribal disorder by claiming the leadership position for himself alone and demanding complete submission.

We are told that Muhammad was born a member of the Banu Hashim clan of the Quraysh, a powerful mercantile tribe that controlled the Arabian trade routes through Mecca. The Quraysh were a typical corporate kin group: subdivided into many clans, the tribe was itself a subdivision of the larger Banu Kinanah tribe. All these clans and tribes were loosely united by their supposed descent from the mythical wanderer Ishmael. This gave them a remote connection to the Jewish descendants of Abraham. It is therefore not an accident that the new Islamic "supertribe" incorporated Abraham and Jesus into its lineage.

The Quraysh rose to prominence when a tribal leader named Qusai ibn Kilab obtained control of the Kaaba, an ancient pagan shrine that attracted numerous pilgrims. This was a lucrative franchise and Qusai ibn Kilab placed family members in control of it, distributing responsibilities (and profits) among the clans of his tribe. Their rivalries continued, however, apparently growing more intense during Muhammad's lifetime.

Muhammad was a religious revolutionary who introduced Abrahamic monotheism into a polytheistic culture. Arabs at that time believed in a supreme deity but also in various lesser gods or tribal deities. Mecca was the center of this polytheistic system. Muhammad's revelation attracted many followers but also drew opposition from powerful clan leaders, whose authority (and income) relied on control of the pilgrimage trade.

In Mecca, Muhammad preached what in today's terms was a religion: prayer to one God, charitable contributions, and the like. The rejection of his message by the polytheists is etched into Islam as a period of persecution of Muslims. To this day, followers of Muhammad's example who encounter the slightest resistance to their preaching speak of persecution.

In 622, these rivals drove Muhammad and his small Muslim community out of Mecca. Muhammad fled to Medina, where he built up his power base through alliances with larger tribes such as the Bakr and Khuza'a. Strategic marriages strengthened his ties with these clans; he himself married the daughters of Abu Bakr and Umar, while Uthman and Ali (Muhammad's cousin) married his daughters. Thus he had family ties with the first four caliphs who succeeded him after his death. During this time Muhammad also promulgated a comprehensive system of moral and political rules, known as the Constitution of Medina, which served to unite the tribes in a community of faith and practice. It was at this point that many tribal practices became an integral part of what evolved to become sharia.

Eight years later, having assembled a large army (known as the Prophet's Companions), Muhammad marched on the Quraysh, who are said to have surrendered without a fight. He then returned to Mecca, married the daughter of the head of the Quraysh, and proceeded to incorporate the other tribes of the Arabian Peninsula into the new Islamic community.

After Muhammad died in 632, a series of lightning conquests by his successors extended Muslim control over an enormous territory—one of the largest empires the world had ever seen. These conquests were extremely brutal and the conquered populations were given a stark choice: convert, die, or (if they were Jews or Christians) accept second-class status as taxpaying *dhimmi*. Most chose conversion and were incorporated wholesale into the growing Muslim supertribe or *ummah*. Yet in many ways the social psychology of Islam remained that of a persecuted tribe, with a powerful "insider/outsider" mentality.

During Muhammad's lifetime, tribal and nationalistic differences within the Islamic community were strongly discouraged. After his death, however, clan rivalries reemerged to shape dynastic struggles in the Caliphate. The Quraysh claimed control and supplied the first three ruling dynasties: the Umayyad, Abbasid, and Fatimid. The Sunni/Shia split was originally a war of succession between two rival lineages—unlike the schisms of Christianity, as we have already noted, it was initially not theological in nature. The passions aroused by this ancient tribal blood feud still divide the Muslim world today.

Medina welcomed Muhammad in part because the local tribal leaders believed their feuding residents might be able to unite around his teachings. Islam would defuse the discord within the city and become a rallying cry against enemies outside. Thus, from the start, Muhammad entered Medina charged not just with spreading his religious message, but also with creating a political order.

The other monotheistic religions were different. The Torah was recorded long after the kingdom of Israel had fallen into ruins. Christian doctrine evolved over centuries, always in the context of a preexisting Roman Empire, one of the strongest polities of the entire premodern period. In Islam, by contrast, the Qur'an was revealed in tandem with its rise and early conquests. In fact, Muhammad's empire began to take shape before all of the verses were compiled in one book. Thus, for Islam, faith and power were from the outset intertwined—indeed inseparable.

Muhammad himself differed in a crucial way from Abraham and Jesus. He was not only a prophet but also a conqueror. He is said to have personally led numerous military campaigns and raiding expeditions. Sahih Muslim, one of the six major authoritative hadith collections, claims he undertook no fewer than nineteen military expeditions, personally fighting in eight of them.[2] Nor did he hesitate to mete out violent reprisals or to enjoy the spoils of war. In the aftermath of the 627 Battle of the Trench, for example, "Muhammad felt free to deal harshly with the Banu Qurayza, executing their men and selling their women and children into slavery."[3] In this way the Prophet became a conquering chieftain. Thus the Qur'an declares, "O Prophet! We have made lawful to thee thy wives to whom thou hast paid their dowers; *and those whom thy right hand possesses [slaves] out of the prisoners of war whom Allah has assigned to thee*" (33:50).[4] (It is, of course, passages such as these that groups like Islamic State or Boko Haram use to justify their actions.)

From a Muslim Reformer's perspective, one of the main problems with Islam is that the tribal military and patriarchal values of its origins were enshrined as spiritual values, to be emulated in perpetuity. The Qur'an emphasizes that all Muslims form one community of believers, the *ummah* (2:143). Although this community superseded prior tribal allegiances, the new religion retained many traditional tribal customs and enshrined them as religious values. These values pertain especially to honor, male guardianship of women, harshness in war, and the death penalty for leaving Islam. As Philip Salzman explains, "Seventh-century Arab tribal culture influenced Islam and its adherents' attitudes toward non-Muslims. Today, the embodiment of Arab culture and tribalism within Islam impacts everything from family relations, to governance, to conflict."[5]

Prior to the rise of Islam, Arab tribes had fought one another, through raiding expeditions and perpetual feuds. Salzman notes that Islam imposed a measure of unity while retaining the traditional tribal habit of the feud "by opposing the Muslim to the infidel, and the *dar al-Islam*, the land of Islam and peace, to the *dar al-harb*, the land of the infidels and conflict."[6] What had been tribal raiding now "became sanctified as an act of religious duty": holy war, or *jihad*.[7] What mattered to Muslims was conquering as much territory as possible and bringing it under Islamic sovereignty, ruled through Islamic holy law.[8]

Muhammad also left behind—true to tribal form—detailed instructions on the division of the bounty gained by Muslim troops through conquest. In Qur'an 8:1 such spoils of war are legitimized. The hadith are full of detailed instructions on what are really norms of tribal conquest. In the authoritative collection Sahih Bukhari alone, there are more than four hundred stories describing military expeditions led by the Prophet Muhammad, and more than eighty stories containing instructions on the appropriate division of booty.[9]

These various residues of tribalism matter because even if Islam is reformed, they are likely to persist. A separation of religion from politics—a distinction between Mecca and Medina—would not do away with the problems created by these inherited tribal norms.

The Honor/Shame Dynamic

Among the most crucial features of the tribal system institutionalized by Islam is the concept of honor. This requires careful explanation for Western readers, whose understanding of terms like "family" and "honor" is fundamentally different. The family structure to keep in mind is an extended kinship group (or clan) whose numbers are increased through practices such as polygamy and child marriage. By having boys marry when they are as young as fifteen or sixteen, the space between generations shrinks, and the number of descendants grows. This kind of family is much like an old *talal* tree, with a deep main root, a solid stem, and myriad branches. Leaves bud, grow, and fall off; branches may be cut and new ones take their place; but the tree stands. Each of its components is dispensable, but the tree itself is not. That is the most important "family value" instilled into children. The individual barely registers in this scheme.

Each person within the kinship group has value to the

tribe as a whole, but certain members are more valuable than others: young men who can go into battle to defend their kin are more useful than young girls or old women. Marriageable girls are more highly valued than older women because they are necessary to produce sons, and can also be traded. Each family's worst nightmare is to be uprooted and destroyed. Given all the possibilities for destruction, the longer a kinship group survives, the stronger it is. Families draw a sense of pride from their history of resilience, passed on through oft-repeated stories and poems about the bloodline.

That pride was what made my grandmother teach me my line of descent back so many generations and hundreds of years. She made it clear to me that it was the duty of young people not only to bask in the inherited glory of their blood-line, but also to maintain it above all else, even if that might cost them their property or their lives. I was also taught to regard anyone outside the bloodline with extreme wariness.

Before Islam was founded, the various extended families of Arabia collaborated and also competed through a network of complex commercial and marital alliances, sometimes allying in battle, sometimes fighting against one another. In this world, conflicts within the clan had to be defused as quickly as possible to preserve the image of strength; infighting would lead to the perception of weakness and make the clan vulnerable to attack. Honor was all-important. Anyone who insulted or humiliated the bloodline must be punished. If one man killed another, for example, the victim's father, brother, uncle, cousin, or son must take revenge, to uphold the clan's honor. And this revenge might be inflicted not just on the killer, but also on his entire family.

Anthropologists since Ruth Benedict's study of Japan in World War II have made a distinction between shame cultures and guilt cultures. In the former, social order is maintained by the inculcation of a sense of honor and shame before the group. If our behavior brings discredit on our tribe, it may punish or even expel us. In a guilt culture, by contrast, a person is taught to discipline himself by means of his own conscience—sometimes backed up by the threat of punishment in the life to come. Most Western societies went through a thousand-year transformation from shame to guilt, a process that coincided with the gradual breakup of tribal family structures. Europeans underwent a long process of detribalization, beginning with subjection to Roman law, conversion to Christianity, the imposition of monarchical rule over baronial power, and the gradual rise of nation-states with their concept of individual citizenship and equality before the law.

The Arab world in which Islam first triumphed did not undergo a similar transition. As Antony Black writes in *The History of Islamic Political Thought*, "Muhammad created a new monotheism fitted to the contemporary needs of tribal society."[10] The effect was to perpetuate tribal norms by freezing them in place as holy writ. Arabs could see themselves as "the chosen people" with "a mission to convert or conquer the world." According to Muhammad, each of the great monotheistic religions was an *ummah*—a community or nation defined by its adherence to the teachings of its prophet. Jews were defined as an *ummah* through their adherence to the book of Moses. Christians were an *ummah* united by adherence to the teachings of the prophet Jesus. The Islamic *ummah*, however, was meant to supersede these other groups. Within the *ummah*, all Muslims were brothers and sisters. Yet this notion did not displace the older ties of the bloodline. As it is set down in the Qur'an: "Blood relations among each other have closer personal ties in the Decree of Allah than (the

Brotherhood of) Believers" (33:6). Despite the rise of a pan-Islamic religious identity in which all individuals notionally submitted to Allah, Islam therefore retained elements of the shame culture.

From its origins as a new faith community, Islam had the overwhelming need to remain unified or risk reverting back to tribal fragmentation. The first schism over the question of succession nearly led to the collapse of the religion. Within Islam, *fitna*—strife or disagreement—was therefore seen as fundamentally destructive. Dissent was a form of betrayal; heresy as well. These individualistic impulses had to be suppressed to preserve the unity of the larger community. Those who wonder at the ferocity of Islamic punishments for dissent fail to grasp the threat that skepticism and critical thinking were believed to pose.

In a clan setting, shameful behavior constitutes a betrayal of the bloodline. In the wider Islamic setting, heresy constitutes a comparable threat, as does outright unbelief—apostasy—both of which are punishable by death. Those who betray the faith must be weeded out to maintain the integrity of the *ummah*.

This belief in the danger of dissent has had powerful consequences, but perhaps the greatest has been to suppress innovation, individualism, and critical thinking within the Muslim world. Muhammad himself, as both the messenger of God and the founder of the Islamic "supertribe," is revered as an irreproachable source of wisdom and a model of behavior for all time. To question his authority in any way is considered an unacceptable affront to the honor of Islam itself.

It is not fashionable today in academic circles to discuss the legacy of Arab clan structures in the development of Islam. It is considered ethnocentric, if not downright orientalist, even to bring it up. But today the Middle East and the wider world are increasingly at the mercy of a combination of the worst traits of a patriarchal tribal society and unreformed Islam. And because of the taboos over what can and cannot be said—taboos backed up by the threat of violent reprisals—we are unable to have an open discussion of these issues.

The Sacrosanctity of the Qur'an

If Muhammad is unique among the prophets, the Qur'an is unprecedented among religious texts. Muslims today are taught that the Qur'an is a complete and final revelation that cannot be changed: it is literally God's last word.

The Qur'an and its related texts are the fundamental source of the Islamic veneration of the afterlife, as well as the call to jihad. They make explicit the concept of commanding right and forbidding wrong and the specific dictates of sharia. In turn, these concepts would not have such enduring power were they not so entwined with the belief in the timeless, all-powerful, and immutable words of Allah and the deeds of Muhammad. Until Islam can do what Judaism and Christianity have done—question, critique, interpret, and ultimately modernize its holy scripture—it cannot free Muslims from a host of anachronistic and at times deadly beliefs and practices.

My first memories of the Qur'an are of my mother and grandmother kissing its cover, of the admonition never to touch it without having first washed my hands, and of sitting on the hot Somali ground as a small child of four or five while the book seemed to tower above us on a high shelf. As I memorized its verses, I was taught simply to obey it. The Qur'an, I learned, was the book sent down "explaining all things" (16:89). It had been revealed to Muhammad by Allah through the Angel Gabriel, beginning when Muhammad lived in Mecca and continuing when he moved to Medina. Gabriel spoke the words one by one to Muhammad, who in turn recited them before scribes. Islamic orthodoxy—not *radical* Islam, but *mainstream* Islamic doctrine—thus insists that the Qur'an is God's own word. Questioning any part of the Qur'an therefore becomes an act of heresy.

The Allah of my childhood was a fiery deity. "On the Day that the enemies of Allah will be gathered together to the Fire," it is written in chapter 41 of the Qur'an, "their hearing, their sight, and their skins will bear witness against them, as to (all) their deeds." Of Abu Lahab, Muhammad's uncle who persistently opposed Islam, it is said in chapter 111: "Burnt soon will he be in a Fire of Blazing Flame! His wife shall carry the (crackling) wood—As fuel!—A twisted rope of palm-leaf fiber round her (own) neck!" Fire is a recurring theme of the Qur'an, and the heat of the desert and the scalding sun, like the crackle of fires at night outside their tents, made these punishments exceedingly vivid to most Arabs, as well as to me. When my mother spoke of "hellfire," she would point to the flaming brazier in our kitchen and tell me: "You think this fire is hot? Now think about hell, where the fire is far, far hotter and it will devour you." The thought gave my sister nightmares. Small wonder I strove to submit to Allah's will.

Later, I learned what it was that made Allah different from the Christian God and Hebrew Yahweh. Allah is not a benevolent father figure, to be depicted in flowing robes with a white beard. In fact, Islam requires that neither God nor Muhammad be depicted in any physical form. Unlike the mosaics of medieval chapels or the frescoes of churches in the Renaissance, every Muslim house of worship from the Grand Mosque down has no human images, only geometric adornments featuring nothing more figurative than enormous flowering plants.

This abstract Allah also reigns supreme as the sole divinity; in Islam there is no Jesus-like son or Holy Ghost. Association of any other god or entity with Allah is considered *shirk* and is one of the gravest sins in Islam—punishable by death according to some scholars. The Qur'an pointedly says, "no son has [Allah] begotten, nor has He a partner in His dominion" (25:2). In Islam, Jesus is recognized as being in the tradition of major Old Testament prophets like Noah and Abraham, but Muhammad is revealed as the last and greatest prophet and the Qur'an is the last word spoken by God. According to Islamic teachings, each prophet up to and including Muhammad opened a window onto the unseen, but after Muhammad's death that window was declared shut until Judgment Day and the end of time. Muhammad was thus the bearer of the last word of God's revelation.[11]

In a similar way, Allah's imperatives for the faithful are not exhortations, such as love thy neighbor, or a covenant, as between God and the Jews, or even a wider moral code, like the Ten Commandments, which address everything from adultery to murder. Rather, first and foremost, Islam commands its followers to perform five religious duties, all of which remind the believers through word and deed that they must above all else submit to the faith and its rules:

1. Have faith in the one God, Allah, and Muhammad, His Prophet;
2. Pray five times a day;
3. Fast during the day for the entire ninth month of Ramadan;
4. Provide charity;

5. Make a pilgrimage to Mecca at least once in a lifetime, if possible.

In its scripture, Islam is also fundamentally different. It places more emphasis on divine omnipotence and less on human free will. "God leads astray whom He will and guides whom He will," it is written. There is even a suggestion in the Qur'an that just as Allah has created what is good, He has also created evil. Chapter 25 says He "created all things, and ordered them in due proportions." This suggests that each person's fate and future have already been established.[12]

Of course, such concepts can also be found in some versions of Christianity. John Calvin was especially insistent on the idea of "double predestination," that God had already chosen who was damned and who saved. The difference is that throughout the history of Christianity there has been intense debate about the relationship between divine omnipotence and human agency. Early debates in Islamic history were eventually won by champions of a heavy determinism, both pertaining to the destiny of one's soul as well as to one's actions in this life.[13] Thereafter, debate on these issues was effectively shut down by zealots who argued that asking such questions was akin to *shirk*, if not to heresy.

Perhaps the biggest problem with the Qur'an's unique status is the fact that the most violent Medina Muslims can find in holy writ justifications for everything they do. Consider the words of Tawfik Hamid, who was once a member of the same radical organization as the Al-Qaeda leader Ayman al-Zawahiri, but is now one of a new generation of Islamic reformers: "The literal understanding of Qur'an 9:29," he has said, "can easily be used to justify what it [Islamic State] is doing. 'Fight those who do not believe in Allah or in the Last

Day and who do not consider unlawful what Allah and His Messenger have made unlawful and who do not adopt the religion of truth from those who were given the Scripture [Jews and Christians]—[fight] until they give the *jizyah* [payment of a tribute tax to Islamic authorities] willingly while they are humiliated.' "[14]

Hamid notes that the four main schools of Islamic jurisprudence agree that this verse means "that Muslims must fight non-Muslims and offer them the following choices: Convert to Islam, pay a humiliating tax called *jizyah* or be killed." Indeed, he adds, "A basic search of almost ALL approved interpretations for the Quran supports the same violent conclusion. The 25 leading approved Quran Interpretations (commentaries)—that are usually used by Muslims to understand the Quran—unambiguously support the violent understanding of the verse."[15]

Hamid's conclusion: while there are certainly many in Islam who are "moderate Muslims," the central truth is that until "leading Islamic scholars provide a peaceful theology that clearly contradicts the violent views of the IS," there will be only a limited space for such moderation.[16]

As the violence committed in the name of Islam is so often justified by the Qur'an, Muslims must be challenged to engage in critical reflection about their most sacred text. This process necessarily begins by acknowledging both its human composition and its numerous internal inconsistencies.

The Qur'an as Text

Muslims have generally shown little interest in subjecting the Qur'an to the same scientific, archaeological, and textual scrutiny the Bible has received.[17] Yet respect for religious beliefs does not require us to suspend our own critical judgment where the Qur'an is concerned, any more than it does in the case of the Old or New Testaments.

Very little is definitely known about the Qur'an's early composition and little work on it was done until quite recently. Western scholars who have studied the Qur'an dispassionately have argued against the traditional Islamic narrative.[18] One of the scholars who took a more critical approach toward early Islamic history was John Wansbrough, who challenged the traditional narrative in two books published in the 1970s, arguing that Islam was originally a Judeo-Christian sect.[19]

Fred Donner, a professor of Near Eastern studies at the University of Chicago, has argued that the Qur'an was originally an orally recited text, and its history in the years following Muhammad's death is "not clear." The survival of various ancient manuscripts indicates that the recitation of the early Qur'anic text "was far from uniform." An early collection of the verses may have been prepared under Caliph Abu Bakr and kept by Caliph Umar, but "it is not clear . . . whether this written collection was complete or not, nor whether it had any official status."[20] An official text is said to have been prepared under Caliph Uthman (644–656), who ordered that competing versions of the Qur'an be destroyed.[21] But in the city of Kufa one of Muhammad's companions, Abdallah Masud, refused Uthman's order. Islamic tradition itself also contains evidence that the Qur'an we know today differs from the original text. The pious Caliph Umar warned Muslims against saying they know the whole Qur'an, because "much of it has disappeared."[22]

Western researchers have advanced several theories about the Qur'an's composition. Günter Lüling believes that it reflects a combination of Christian texts that have been given a new Islamic meaning, and "original Islamic passages which had been added to the Christian ones." For Lüling, the Qur'an is a composite work shaped by human hands and human editors. Gerd Puin's study of ancient manuscripts found in Yemen led him to conclude that the Qur'an is a "cocktail of texts," some of which may have predated Muhammad by a century.[23] Christoph Luxenberg (a pseudonymous scholar) theorizes on the basis of linguistic analysis that there exists a gap of one and a half centuries between the Qur'an's first publication and the final editing process through which it received its traditional form.[24] Fred Donner suggests another possibility: it may be a composite of different religious texts from various communities in Arabia. Certainly, there are significant variations in spelling in different versions of the Qur'an.[25]

What might have motivated people to compile a document like the Qur'an? Malise Ruthven offers the "revisionist theory":

> that the religious institutions [of Islam] emerged at least two centuries after Muhammad's time, to consolidate ideologically, as it were, the Arab conquest. [This theory] would mean that the Arabs, anxious to avoid becoming absorbed by the more advanced religions and cultures of the peoples they conquered, *cast about for a religion that would help them to maintain their identity.* In so doing they looked back to the figure of the Arabian Prophet, and attributed to him the reaffirmation of an ancient Mosaic code of law for the Arabs.[26]

Ruthven notes that the revisionist theory, if true, would help explain why the *qiblas* of certain early mosques in Iraq face Jerusalem rather than Mecca.[27] Other evidence indirectly supports this theory of later authorship. Tarek Fatah, founder of the Muslim Canadian Congress, has argued that a story about Muhammad—in which a Jewish tribe surrendered to the Islamic army in the city of Medina and the Prophet personally beheaded between six hundred and eight hundred prisoners of war—may in fact be a creation of later Muslim rulers, two hundred years after the incident was said to have occurred (627 CE). (This story is not in the Qur'an, but it shows how easily the life of the Prophet could be embroidered long after the fact.)

It is, to say the least, difficult in the face of all this evidence to deny that there was a human influence involved in composing what is now known as the Qur'an. Yet Islamic thinkers such as the late Pakistani Abul A'la Mawdudi have declared without hesitation that the Qur'an "exists exactly as it had been revealed to the Prophet; not a word—nay, not a dot of it—has been changed."[28] And that remains mainstream Muslim doctrine.

All scriptures contain contradictions and the Qur'an is no exception. But Islam is the only religion that has promulgated a doctrine to reconcile the Qur'an's contradictions in order to maintain the belief that it is the direct revelation of God. As Raymond Ibrahim observes:

> No careful reader will remain unaware of the many contradictory verses in the Quran, most specifically the way in which peaceful and tolerant verses lie almost side by side with violent and intolerant ones. The ulema were initially baffled as to which verses to codify into the Shari'a worldview—the one that states there is no coercion in religion (2:256), or the ones that command

believers to fight all non-Muslims till they either convert, or at least submit, to Islam (8:39, 9:5, 9:29).[29]

To explain these contradictions, Islamic scholars developed a doctrine known as "abrogation" (*an-Nasikh wa'l Mansukh*), whereby Allah issues new revelations that supersede old ones.

Take, for example, the specific injunctions regarding war and peace. These successive revelations follow a distinctive arc in the course of the book: they begin in the early "Mecca" sections with admonitions of passivity in the face of aggression; then they give permission to fight back against aggressors; then they exhort Muslims to fight aggressors; finally, Muslims are commanded to fight all non-Muslims, whether they are the aggressors or not. What explains this pattern of gradually increasing aggressiveness? Most likely, it is the growing power and strength of the early Islamic community. Yet orthodox Muslim scholars insist that these changes have nothing to do with contingent circumstances.

Thus Ibn Salama (d. 1020) argued that chapter 9, verse 5, known as *ayat as-sayf*, or the sword verses, abrogated some 124 of the more peaceful Meccan verses.[30] The same applies to the verses concerning forcible conversion. As Ibrahim explains, "whereas Allah supposedly told the prophet that 'there is no compulsion in religion' (2:256), *once the messenger grew strong enough*, Allah issued new revelations calling for all-out war/jihad till Islam became supreme (8:39, 9:5, 9:29, etc.)."[31]

Mainstream Islamic jurisprudence continues to hold that the sword verses (9:5 and also 9:29) have "abrogated, canceled, and replaced" those verses that call for "tolerance, compassion, and peace."[32] This same doctrine is also applied to apparent flaws or contradictions in Muhammad's personal behavior. Suggesting, for example, that Muhammad chose to break a treaty with the Quraysh, rather than being provoked by their dishonorable behavior, has led to threats and violence against Western scholars and journalists. The goal in each instance is to place the Qur'an beyond criticism and reproach. After all, how can one argue with God's word?

Of course, the Qur'an is not the only Islamic text. Accompanying it is the Hadith, the record of Muhammad's sayings, the customs he followed, his teachings, and the personal examples that he left for all Muslims to follow, as well as assorted commentaries on his life. These texts were supposedly written or dictated by those who knew him, including his original companions and his wives. We have every reason to want to know more about the provenance and human composition of these texts, too. But the main questions that have been raised relate to the Qur'an. These include:

- What did the Qur'an retain (or copy) from previous Jewish and Christian holy texts?
- What was Muhammad's contribution to the text now known as the Qur'an?
- Which other individuals (or groups) composed the Qur'an?
- What was added to the Qur'anic draft after the death of Muhammad?
- What was edited out or rephrased from the original Qur'an?

The answers to some of these questions may never be fully known, but we have a duty to ask them—and to protect the lives and liberty of those grappling with them, Muslims and non-Muslims alike.

Leading the effort to bring modern methods to the study of the Qur'an is Professor Angelika Neuwirth of the Free University in Berlin. The research program she leads, Corpus Coranicum, is housed at the Brandenburg Academy of Science and Humanities and will likely take decades to complete.[33] But analyzing the Qur'an is not like studying the holy texts of Judaism or Christianity. When two German researchers traveled to Yemen to take pictures of old Qur'anic manuscripts, the authorities confiscated the pictures. Although diplomats eventually secured the release of most of the pictures, the episode sparked predictable reactions. One letter to the *Yemen Times* read: "Please ensure that these scholars are not given further access to the documents. Allah, help us against our enemies."[34]

The language of the Qur'an is Arabic, and to many Muslims that remains the divine language. To this day there are tremendous disputes about whether it is acceptable to translate it into other languages. That is partly because, unlike the Bible, the Qur'an is supposed to be learned by heart. As the Islamic scholar Michael Cook puts it, "The Muslim worshiper does not *read* the Qur'an, but rather *recites* it." All 77,000 words, roughly 6,200 verses, of the Qur'an must be internalized, giving it what Cook calls "a degree of scriptural saturation of daily life which is hard for most inhabitants of the Western world to imagine."[35] In early-nineteenth-century Cairo, for example, parties and gatherings held by the city's middle and upper classes often featured a recital of the Qur'an, usually by three or four trained reciters, spanning as many as nine hours. Guests might come and go, but the recitation of the verses was continuous.

This highlights another important difference with other monotheistic scriptures. Although the Qur'an makes reference to some stories found in both the Torah and the Bible, it is distinctly not a storytelling text; no sustained meta-narrative binds it together. The Qur'an is not designed to be read as literature. Nor can scenes from it be depicted as scenes from the Bible were in works of art like Michelangelo's Sistine Chapel or Leonardo's *Last Supper.* It does not have multiple narrators, like the Bible, but rather relies on one voice throughout, which the reciter is essentially channeling.

It is hard to convey to a non-Muslim how the recitation of the Qur'an embeds the text socially. In the middle of the twentieth century, for example, ordinary Egyptians riding public trams would move their lips, silently mouthing scripture as they traveled from stop to stop.[36] I can well remember how when someone in my family lay sick or dying—like my aunt when she contracted breast cancer—the Qur'an was chanted by the bedside, in the belief that its words alone would cure the patient. Analogies with Christian prayer are misleading because the reciter of the Qur'an is voicing God's words, not appealing to God for intercession.

Does the Qur'an Inspire Violence?

If the Qur'an were used only to heal the sick, there would be less need for a Muslim Reformation. Unfortunately, as we have seen, it is also very commonly cited today to justify acts of violence, including all-out war against the infidel.

David Cook, a professor of religious studies at Rice University who has carefully studied jihad, notes that in the Qur'an, "the *root* (the verbal derivatives) of the word *jihad* appears quite frequently with regard to fighting (e.g., 2:218, 3:143, 8:72, 74–75, 9:16, 20, 41, 86, 61:11) or fighters (mujahidin, 4:95, 47:31)."[37] Most verses in the Qur'an, Cook emphasizes, "are unambiguous as to the nature of the jihad prescribed—the vast majority of them referring to 'those who believe, emigrate, and fight in the path of Allah.' "[38] In the historical evolution of Islam, "the armed struggle—aggressive conquest—came first, and then additional meanings became attached to the term [jihad]."[39]

To be sure, there are stories of violence and brutality in the Torah and Bible. When King David's daughter, Tamar, is raped by her half-brother, David imposes no punishment and Tamar is discarded and shamed. But Talmudic and biblical scholars today do not sanction sibling rape. Instead, they are most likely to express grief for Tamar and revulsion at the crime, and to show how this one act led to the unraveling of David's family. Contrast this with the use by modern Islamic scholars of Muhammad's decision to marry a six-year-old girl, consummating their marriage when she turned nine, to justify child marriage in Iraq and Yemen today.

The literal reading of the Qur'an is a central part of what animates the bloody battles of jihad playing out across Syria and Iraq. Many of today's Sunni and Shiite fighters believe they are participating in battles foretold in seventh-century prophecies—the accounts in the hadith that refer to the confrontation of two massive armies in Syria. "If you think all these mujahideen came from across the world to fight Assad, you're mistaken," a Sunni Muslim jihadist who uses the name Abu Omar explained to a Reuters reporter in 2014. "They are all here as promised by the Prophet. This is the war he promised—it is the Grand Battle."[40] "We have here mujahideen from Russia, America, the Philippines, China, Germany, Belgium, Sudan, India, and Yemen and other places," a journalist was told by Sami, a Sunni rebel fighter in northern Syria. "They are here because this is what the Prophet said and promised, the Grand Battle is happening."[41] In much the same way, the leader of Boko Haram cites the Qur'an as his excuse to sell 276 kidnapped Nigerian schoolgirls into slavery.

Reason and the Qur'an

If Muhammad and the Qur'an are providing justifications for so much wrongdoing in the world, then it must be of more than scholarly interest to apply the tools of reason to both Prophet and text. The problem is that Islamic scholars arguing in favor of human reason have long been on the losing end of doctrinal conflicts. When rationalists squared off against literalists during the seventh, eighth, and ninth centuries, they lost. The rationalists wanted to include in Islamic doctrine only principles based on reason. The traditionalists countered that the human intellect is "defective, fickle, and malleable."[42]

Changing central aspects of Islamic doctrine became even more difficult in the tenth century. At that time, jurists of the various schools of law decided that all the essential questions had been settled and that permitting any new interpretations would not be productive. This famous episode is referred to as the closing of "the gates of *ijtihad*." The gates of reinterpretation were not suddenly slammed shut: it was a gradual process. But once shut, they proved impossible to reopen. The late Christina Phelps Harris of Stanford University summarized the impact as creating "a framework of inexorable legal rigidity."[43]

In this process a key role was played by the imam Abu Hamid Muhammad ibn Muhammad al-Ghazali, who died in AD 1111. Al-Ghazali detested the ancient Greek philosophers. He regarded human reason as a cancer upon Islam. His most famous work is *Incoherence of the Philosophers*, which attacks and refutes the claims of the ancients. Against their pretensions, al-Ghazali posits an all-knowing God. Allah knows the smallest particle in heaven and on earth. And because Allah knows everything and is responsible for everything, he already knows and has fully formed every part of the world and every action, from whether an arrow reaches its target to whether a hand is waved. Thus, al-Ghazali writes, "Blind obedience to God is the best evidence of our Islam." Those, such as the Andalusian scholar Ibn Rushd, who disagreed with al-Ghazali found themselves exiled, or worse.

Nine hundred years have passed, and yet al-Ghazali is still considered by many in Islam to be second only to Muhammad. He provided the standard answer to almost any question posed in Arabic: "Inshallah," meaning "If Allah wills it" or "God willing." The latest flowering of al-Ghazali's concepts can be found today in the teachings of groups such as Boko Haram (whose very name means "Non-Muslim teaching is forbidden"), Islamic State, and Southeast Asia's Jemaah Islamiyah. They adhere to the principle of "*al-fikr kufr*," that the very act of thinking (and along with thinking, education, reason, and knowledge) makes one an infidel (*kufr*). Or as Taliban religious police have written on their propaganda placards: "Throw reason to the dogs—it stinks of corruption."[44]

There is in fact no good reason al-Ghazali and his ilk should have the last word in defining Islam. Muslims around the world cannot go on claiming that "true" Islam has somehow been "hijacked" by a group of extremists. Instead they must acknowledge that inducements to violence lie at the root of their own most sacred texts, and take responsibility for actively redefining their faith.

The crucial first step in this process of modification will be to acknowledge the humanity of the Prophet himself and the role of human beings in creating Islam's sacred texts. When Muslims tell us that the Qur'an is the immutable and unchanging word of God, that it is entirely consistent and infallible, and that none of its injunctions and commandments can be treated as in any way optional for true believers, we need to retort that, by the lights of scholarship and science, this is simply not the case. In truth, Islamic doctrine is adaptable; certain parts of the Qur'an were abrogated over time. So there is no reason to insist that the militant verses of the Medina period should always be given priority. If Muslims wish their religion to be a religion of peace, all they have to do is "abrogate" those Medinan verses. Mahmoud Muhammad Taha, who was executed in 1985 for "apostasy" in Sudan, proposed to do just that.[45]

The next step in dismantling the ideological foundation of Islamist violence will be to persuade Muslims raised on an alluring vision of the afterlife to embrace life in this world, rather than actively seeking death as a path to the next.

CHAPTER 4

THOSE WHO LOVE DEATH

Islam's Fatal Focus on the Afterlife

On October 4, 2014, inside Chicago's O'Hare Airport, three American-born teenagers were apprehended by the FBI. The two brothers, aged nineteen and sixteen, and their seventeen-year-old sister were on their way to Turkey, where they planned to cross the border into Syria and join Islamic State. The three left behind letters for their parents, devout Muslims who had immigrated to the United States from India. The eldest, Mohammed Hamzah Khan, explained that "Muslims have been crushed under foot for too long," adding that the United States is "openly against Islam and Muslims," and that he did "not want my progeny to be raised in a filthy environment like this."[1]

But the sister took a different tack. She wrote to her parents: "Death is inevitable, and all of the times we enjoyed will not matter as we lay on our death beds. Death is an appointment, and we cannot delay or postpone, and what we did to prepare for our death is what will matter." In a striking irony, the girl who wrote those lines celebrating the primacy of death was planning to become a physician.

Like her brothers, she had attended a private Islamic school for nearly all her educational life. There she had demonstrated the highest facility with the Qur'an, becoming "Hafiz," meaning that she had memorized the entire text in Arabic.

In short, the decision of these siblings to join IS was not the result of knowing too little about Islam, much less of ignorance of the sacred texts. Nor can we ascribe their choice to poverty, social deprivation, or limited opportunity. The family lived in a comfortable Chicago suburb, the children attended private school, they had computers and cell phones—although, in a classic example of cocooning, the parents got rid of their television when their eldest child was eight because they wanted to "preserve their innocence."

Rather, this was a choice directly underpinned by contemporary Islamic philosophy and, in particular, its contempt for many of the central values of the West. In the words of a local Islamic community leader, Omer Mozaffar, who teaches theology at the University of Chicago and Loyola University Chicago, Muslim parents "think 'American' equals 'immoral.'"[2]

And it is not simply our American shopping malls, chain restaurants, movies, and music downloads. It is our values, our social fabric, our very way of life. Americans are raised to believe in life, liberty, and the pursuit of happiness. Muslims such as the Chicago Three, by contrast, are educated to venerate death over life—to value the promise of eternal life more highly than actual life here on earth. They see their primary purpose in this life as preparing for death: in the words of that Chicago teenager, "what we did to prepare for our death is what will matter."[3] Death is the goal, the event that matters because it leads to the prize of eternal life.

Many Muslims today believe this with a fervor that is very hard for modernized Westerners to comprehend. By contrast, the leaders of IS and similar organizations know exactly how to exploit the Islamic exaltation of death—to the extent that three American teenagers would spend $2,600 on plane tickets with the ultimate goal of hastening their own deaths.

Life and Afterlife

The afterlife is as central to the Islamic mind as the clock has become to the Western mind. In the West, we structure our lives according to the passage of time, what we will accomplish in the next hour, the next day, the next year. We plan according to time and we generally assume that our lives will be long. Indeed, I have heard Westerners in their eighties talking confidently as if they have decades still to live. The old Christian preoccupations with mortality—so vividly expressed in Shakespeare's *Hamlet* or in the poetry of John Donne—have receded in the face of rising life expectancy, actuarial calculation, and increasingly secular thinking. In the Islamic mind, by contrast, it is not the ticking of the clock that is heard, but the approach of the Day of Judgment. Have we prepared sufficiently for the life that will come after death?

The problem before us, then, is not simply one of better

education: the people who hold this belief are not ignorant laborers but highly educated and skilled engineers and doctors. Focusing on death is what they are taught from the beginning of their lives. It was what I was taught from the beginning of mine.

From the time I could learn the most basic lessons, I was taught that our life on this earth is short and that it is temporary. During my childhood, countless people died: relatives died, neighbors died, strangers died—from disease, from malnutrition, from violence, from oppression. Death was on our lips all the time. We got so used to it and it became such a part of us that we wouldn't speak without mentioning it. I could not make the simplest plans with a friend without saying, "See you tomorrow, if I'm alive!" or "If Allah wills it." And the words made perfect sense because I knew that I could die at any time.

I was also told that all of your life is a test. To pass that test, you must follow a series of obligations and abstain from all that is forbidden, so that when it comes to the final trial of judgment before Allah, you will be admitted to paradise, an actual place with water and date trees heavy with fruit. Thus, from the beginning, as a Muslim child, I was taught to invest my actions, my thoughts, my creativity not in the here and now, but in the hereafter. The ultimate lesson I learned was that your real, eternal life starts only after you die.

I believed all of this without question—until I reached Holland. There no one talked about death, let alone life after death. Without equivocation they said, "See you tomorrow!" And if I replied, "If I'm alive!" they would look at me quizzically and say, "Of course you'll be alive. Why ever not?"

Martyrdom vs. Sacrifice

What are the origins of the Muslim cult of martyrdom? After Muhammad's *hegira* to Medina, he and his small armies faced far larger, more powerful forces. Both the Qur'an and the hadith describe how Muhammad and his cohorts defeated them because Allah was on their side. Allah blessed their wars as jihad—holy war—and declared that the most glorious Muslim warriors were the *shaheed*, the martyrs. So the men on the field not only welcomed war, they welcomed death in war because it elevated their status in paradise.

The belief that this life is transitory and that it is the next one that matters is one of the core teachings of the Qur'an. For the believer looking to find glory in death, there are numerous passages like this: "Only he who is saved far from the Fire and admitted to the Garden will have attained the object (of Life): For the life of this world is but goods and chattels of deception" (3:185). Elsewhere, the Qur'an emphasizes the transitory nature of the world. "Thou seest the mountains and think them firmly fixed, but they shall pass away as the clouds pass away" (27:88). Everything on earth is temporary; only Allah is permanent.

Such is the importance of martyrdom in Islam that martyrs have all their sins forgiven and automatically ascend to the highest of the seven levels of paradise. One sentence in the *Princeton Encyclopedia of Islamic Political Thought* drily captures this concept. After burying martyrs, usually in the clothes in which they had fought, "most jurists were of the opinion that there was no need to say the funerary prayers over the martyr's body, the assumption being that all his sins had been forgiven and that he would ascend to heaven right away."[4]

The Qur'an includes a very vivid depiction of paradise for the believing, repentant Muslim, far more precise than any visions of heaven in Christianity or the even more nebulous versions of a possible hereafter in Judaism:

> There will be two Gardens containing all kinds (of trees and delights); In them (each) will be two Springs flowing; In them will be Fruits of every kind, two and two. The Fruit of the Gardens will be near (and easy of reach). In them will be (Maidens), chaste, restraining their glances, whom no man or Jinn before them has touched; Like unto Rubies and coral. Is there any Reward for Good—other than Good? (55:46–60)

As if that were not detailed enough, here is a hadith narrated by the famous scholar al-Ghazzali:

> These places [in paradise] are built of emeralds and jewels and in each building there will be seventy rooms of red color and in each room seventy sub-rooms of green color and in each sub-room there will be one throne and over each throne seventy beds of varied colors and on each bed a girl having sweet black eyes. . . . There will be seven girls in each room. . . . Each believer will be given such strength in the morning as he can cohabit with them.[5]

These virgins "do not sleep, do not get pregnant, do not menstruate, spit, or blow their noses, and are never sick."[6]

Significantly, there is relatively little in this Qur'anic discussion of paradise for women. It is also unclear whether a woman's paradise is the same as a man's, or what a woman's paradise might be like. Even in death, there is an assumption that a woman is less than a man. Nouman Ali Khan, who is listed by the Royal Islamic Strategic Studies Centre in Amman, Jordan, as one of the world's five hundred most influential Muslims, is a very Westernized (and very glib) cleric who also heads the Bayyinah Institute in Dallas. Wearing a crisp blue dress shirt, he explains on YouTube that, once in heaven with Allah, all of a wife's annoying traits are removed. "So don't get depressed," he says, joking that when you first encounter your wife, you will say, "So you're here too? I thought this was . . ." Only in *jannah*, in paradise, does your wife have the traits that you actually want.

For Christians, heaven is simply a place without suffering, a place of peace. The precise nature of that peace is seldom spelled out. For Muslims, by contrast, paradise is a goal, a destination, a place infinitely preferable to the one where we reside. "Dear wise brother," says the Egyptian imam Sheikh Muhammad Hassan in an online sermon, "your real life starts with your death, and so does mine."[7]

How exactly does the preeminence of the hereafter get drummed into Muslims? To start with, it is invoked five times a day in ritual prayer. Then there are the constant reminders. The next life is the life that matters, not this one, you are told. You will not please God by going to your job and working hard. You will please God by spending more time praying, more time proselytizing, by fasting during Ramadan, by journeying to Mecca. You can be redeemed, you can salvage whatever you have lost, not by devoting yourself to improving your life in the here and now, but by following religious dictates and achieving entry into paradise. And the most spectacular way to enter paradise is as a martyr, by the open embrace of an early death.

In as many generations as my Somali grandmother taught me to count back in our lineage, Islamic conceptions of the afterlife have remained remarkably fixed. Death in holy war and martyrdom continue to be the most hallowed pathway to paradise. The Enlightenment, evolution, Einstein: none has modified the overarching Islamic vision of paradise or hell, nor its centrality in Islamic theology.

Sacrifice in the Non-Muslim World

Of course, other religions have the concept of an afterlife. Christianity, too, has a tradition of venerating martyrs. John Foxe's 1563 *Book of Martyrs* was one of the most popular publications of the English Reformation. Yet there are important differences in the way the other monotheistic faiths now understand both concepts.

Of the three great religions, Judaism has the least comprehensive concept of the afterlife. Indeed, early biblical writings say very little about what happens after death. When an individual transgresses in the Torah, God punishes the wrongdoer or his descendants in this life. Unlike either Christianity or Islam, Judaism did not see violent death as something that would bring a person closer to God. Over time, some strands of Judaism developed a clearer conception of an afterlife, but in the wake of the Holocaust, many Jews have returned to the religion's original conceptions, seeing life on earth as the primary focal point.

Christianity, by contrast, has the idea of heaven at its very heart. That there is life after death is at the very core of Jesus's teaching. He himself demonstrated that with his own Resurrection after his death on the cross. For believers, entrance to the kingdom of God was not based on status—indeed, according to Jesus, the most lowly would be first in line: the poor, the ignorant, the young. Admission was based on being pure of heart, on loving one's neighbor as oneself. People who hoped to enter the kingdom—the "godly"—had to behave on earth toward one another as if they were already there. Persecution of the early Christians encouraged an enduring cult of martyrdom, to be sure. But unlike Muslim martyrs, Christian martyrs were nearly always the unarmed victims of cruel executions, a select few of them attaining sainthood precisely because of their sublime sufferings.

Unlike Islam, Christianity has never been a static religion. A three-tiered universe features in much medieval iconography, with heaven on top, earth in the middle, and hell below. That was later modified to include Purgatory, a kind of waiting room for those who had not fully atoned for their sins on earth and must endure additional purgation before being admitted to heaven. As we have seen, the Reformation was initially a revolt against the Catholic Church's practice of selling shortcuts out of Purgatory. But it was not a revolt against the notion of an afterlife. On the contrary: the wars of religion that raged in Europe from the 1520s to the 1640s saw a revival of the early Church's cult of martyrdom. As Catholics and Protestants burned each other alive, the list of Christian martyrs grew steadily longer. And the more wars Christians fought—whether against one another or against "heathens" abroad—the more the ideal of the warrior martyr took hold. Christianity and Islam never resembled each other more closely than in their periodic military collisions, from the Crusades onward.

Today, in our age of space travel and deep drilling beneath the earth's surface, it has become difficult to maintain a literal conception of an actual heaven above and a hell below. Scientific and medical advances have radically modified the Christian conception of the afterlife, rendering it metaphorical for many believers. To be sure, there are still many Christians who regard the Bible as a factual account of the history of the world from the Creation to the Resurrection. But there are at least as many for whom it is a largely allegorical work, the spiritual meaning of which transcends the acts, miraculous and otherwise, that it purports to record.[8] There are sincere and reputable people on both sides. They disagree, but their disagreement has not undone Christianity. And neither side is blowing anyone else up over it. Week in and week out, rabbis, ministers, and priests do not stand before their congregations, preaching about the world to come and exhorting them to seek martyrdom as a fast track to heaven. Bereaved Christians still seek solace in the thought that they will be reunited with lost loved ones in the hereafter, but no priest today would urge his flock actively to seek death for themselves and others in order to receive a posthumous reward. Murder and suicide are proscribed, not encouraged.

Indeed, most Jews and Christians today recoil from the notion of human sacrifice. For example, most modern believers are deeply uncomfortable with the story of Abraham's attempt to sacrifice his son Isaac to appease God. What has persisted in the Judeo-Christian world is the concept of self-sacrifice as a noble act *when it aims to preserve the lives of others.* In the United States, we expect the men and women of our armed forces to be willing to die to protect their fellow citizens. The president and Congress award the Medal of Honor to military personnel who have taken heroic actions to save others.

If you want to understand the completely irreconcilable

difference I am talking about, you need only compare two groups of people: the perpetrators of the 9/11 attacks, flying their hijacked planes into the World Trade Center, and the New York City firefighters running up the stairs of the burning Twin Towers, determined to save whoever they could, regardless of the risk to their own lives. The West has a tradition of risking death in the hope of saving life. Islam teaches that there is nothing so glorious as taking an infidel's life—and so much the better if the act of murder costs you your own life.

Martyrdom and Murder

As we have seen, Islam is not unusual in having a tradition of martyrs. What is unique to Islam is the tradition of murderous martyrdom, in which the individual martyr simultaneously commits suicide and kills others for religious reasons.

The first modern "martyrdom operation" was in fact inflicted on the perpetrator's fellow Muslims.[9] It was carried out in November 1980 by a thirteen-year-old Iranian boy who strapped explosives to his chest and blew himself up underneath an Iraqi tank during the early part of the Iran-Iraq War. Iran's Ayatollah Khomeini immediately declared the boy a national hero, as well as an inspiration for other volunteers to sacrifice themselves. And in the intervening years, such martyrs have stepped up by the thousands. Suicide bombing remains one of the most common ways in which Shia and Sunni Muslims kill each other.

Another early martyrdom operation was the 1983 suicide bombing of the U.S. Marine Corps barracks in Lebanon, which left 241 American military personnel dead. The attack, conducted by members of a then-obscure group called Islamic

Jihad, so shocked the American public that President Reagan ordered the immediate withdrawal of U.S. troops, handing the jihadists a prestigious victory and confirming the tactic's effectiveness. Since then, Palestinian militants have used suicide bombings repeatedly against Israeli targets. After the U.S. invasion of Iraq, suicide bombings became a recurrent feature of an insurgency that rapidly took on the character of a Sunni-Shia civil war. Suicide bombings are now commonplace events all over the Muslim world, from Afghanistan and Pakistan to Nigeria.

The psychology of suicide bombing is complex. Muslim clerics take great pains to reject the term "suicide," preferring "martyrdom." Suicide, they explain, is for those without hope. Martyrs are living successful lives, but nobly choose to sacrifice their lives for the higher good. These purveyors of death are recognized and honored as well. Within the Palestinian territories, streets and squares are named for them. Mothers of suicide bombers talk as if their sons had gone off to get married. This is not a strange, inexplicable failure of parental love, as some Westerners might like to believe. It is part of an alternative ideology. In this ideology, death is—to quote the seventeen-year-old would-be martyr from Chicago—"an appointment" that must be kept.[10]

True, while the martyrs' ultimate goal might be paradise, for years there were also significant monetary incentives for suicide bombers. The Iraqi dictator Saddam Hussein openly paid the families of Palestinian suicide bombers up to $25,000 for attacks on Israelis. Officials from the Arab Liberation Front would personally deliver the checks, with the compliments of Baghdad.[11] In addition, charities from Saudi Arabia and Qatar have sent money to the families of Palestinians killed in operations against Israel.

Yet it is impossible to explain the cult of murderous martyrdom purely in these material terms. The parents of the 9/11 attackers were not enriched by their sons' bloody deed. In very few societies can it truly make economic sense for a young person—in whom a family must have invested at least a childhood's worth of food, clothing, shelter, and education—to self-destruct.

In the aftermath of 9/11—to date, the most spectacular martyrdom operation ever undertaken—American commentators debated whether the terrorists who flew the hijacked planes into the World Trade Center were "cowards" for attacking a civilian target. Elsewhere, anti-Americans of every stripe hailed the terrorists as heroes. In fact they were neither cowards nor heroes—they were religious zealots acting under the deluded belief that they would not suffer at all as the planes collided with the towers, but would go directly to paradise. You cannot call someone a coward who does not fear death but rather longs for it as an express ticket to heaven. Indeed, you cannot define them at all using the usual Western terminology.

Modern Martyrdom

Today the call to martyrdom can be heard not just in mosques, but also in schools and in the electronic media, from television to YouTube. The argument is a subtle one that is not well understood in the West. During an interview on Al-Aqsa television in May 2014, Dr. Subhi Al-Yazji of the Islamic University in Gaza acknowledged, "the Islamic concept of sacrifice motivates many of our youth to carry out martyrdom operations." But he added:

> Contrary to how they are portrayed by the West and some biased media outlets, which claim that they are youths of eighteen to twenty years who have been brainwashed, most of the people who sacrificed their lives for the sake of Allah were engineers and had office jobs. They were all mature and rational. Some people claim that they did this for the money. [But] take, for example, someone like brother Sa'd, who was an engineer, had an office job, owned a home and a car, and was married—what made him embark on jihad? He believed that the Muslim faith requires us to make sacrifices.[12]

Ismail Radwan, an Islamic University professor and spokesman for Hamas in Gaza, explains what the reward will be for those who embrace death. "When the Shahid (Martyr for Allah) meets the Lord," he writes, "all his sins are forgiven from the first gush of blood, and he is exempted from the torments of the grave. He sees his place in Paradise. He is shielded from the Great Shock and marries 72 Dark-Eyed [Virgins]. He is a heavenly advocate for 70 members of his family. On his head is placed a crown of honor, one stone of which is worth more than all there is in this world."[13]

In part because the Palestinians have been the most frequent proponents and practitioners of suicide bombing, they have developed the most elaborate and detailed rationalizations of martyrdom. To many of them, the afterlife is not a theoretical, abstract concept; it is exceedingly real.[14] As the Tel Aviv disco bomber explained in his will, written before his June 2001 attack, which left twenty-three Israeli teenagers dead, "I will turn my body into bombs that will hunt the sons of Zion, blast them and burn their remains. . . . Call out in joy, oh mother! Distribute sweets, oh father and brothers!

A wedding with the black-eyed [virgins] awaits your son in Paradise."[15]

As a mother of a three-year-old son, I can imagine nothing more unbearable than his death. So I have tried hard to understand the psychology of Mariam Farhat, the Palestinian "mother of martyrs" also known as Umm Nidal, who positively encouraged three of her sons to undertake attacks on Israel that cost them their lives. "It is true that there is nothing more precious than children," she said before one of her sons died in a suicide attack she herself had planned, "but for the sake of Allah, what is precious becomes cheap."[16] Her son Muhammad Farhat attacked an Israeli settlement school with guns and hand grenades, killing five students and wounding twenty-three others before being killed himself. Why did she condone this? "Because I love my son," she replied, "and I wanted to choose the best for him, and the best is not life in this world":

> For us there is an Afterlife, the eternal bliss. So if I love my son, I'll choose eternal bliss for him. As much as my living children honor me, it will not be like the honor the Martyr showed me. He will be the intercessor on the Day of Resurrection. What more can I ask for? Allah willing, the Lord will promise us Paradise, that's the best I can hope for. The greatest honor [my son] showed me was his Martyrdom.[17]

The Palestinian academic Sari Nusseibeh commented that Nidal's words made him "recall the words of the hadith that 'Paradise lies under the feet of the mothers.'"

As the organization Palestinian Media Watch explains, this message "comes from all parts of society, including religious leaders, TV news reports, schoolbooks, and even music videos. Newspapers routinely describe the death and funerals of terrorists as their 'wedding'. . . . The longest running music video on PA TV, originally aired in 2000 and broadcast regularly in 2010, shows a male martyr being greeted in Islam's Paradise by dark eyed women all dressed in white."[18] Yet this cult of murderous martyrdom is no longer confined to the Palestinians. It is not only in Gaza that kindergartners are dressed up as suicide bombers. All across the Muslim world, children are being inculcated with a death wish. On Egyptian television, the child preacher Abd al-Fattah Marwan extolls "the love of martyrdom for the sake of Allah." On Al-Jazeera, a ten-year-old Yemeni boy chants a poem he has composed himself, promising, "I will become a martyr for my land and my honor."[19]

In Somalia, fathers recruit their children, some as young as ten, to become suicide bombers and film their "martyrdom operations" with the same pride as an American father filming his son scoring a goal or hitting a home run. The leaders of Boko Haram likewise raise their children to be martyrs.[20] Finally, and inevitably, the cult of death has reached European Muslims. In 2014 a British-born woman calling herself Umm Layth tweeted a breathless comment on her new life as the wife of a Syrian IS fighter: "Allahu Akbar, there's no way to describe the feeling of sitting with the Akhawat [sisters] waiting on news of whose Husband has attained Shahada [in this case meaning martyrdom]."[21] At the time she wrote those words, Umm Layth had more than two thousand Twitter followers.

Such ideas are already established in America. Consider the very popular *Methodology of Dawah el-Allah in American Perspective*, by Shamim Siddiqi, a leading commentator on Muslim issues, and published by the Forum for Islamic Work. The book sets out how Muslims can establish an Islamic state in the United States and more broadly in the West. It presents both the preferred ways of reaching potential adherents—through mosques, conferences, television and radio appearances—and the best strategies for doing so. But what is most striking is the book's death-laden language, starting in its very first pages. It is dedicated to those "who are struggling and waiting to lay down their lives for establishing God's Kingdom on earth" and quotes the Qur'an on its dedication page: "Of the believers are men who are true to that which they covenanted with ALLAH. Some of them have paid their vow by death (in battle), and some of them still are waiting; and they have not altered in the least" (33:23). Siddiqi focuses on how the ideal Muslim must sacrifice everything for the sake of the Islamic movement and "expect rewards from Allah only in the life hereafter." The perfect Muslim "prefers to live and die for [the hereafter]. He gladly gives up his life for its sake. . . ."[22] Unfortunately, this isn't mere rhetoric.

Fatalism in This World

I can already hear the complaints: Oh, but you are merely citing the extremes; the overwhelming majority of Muslims are not sending their children off to die. And no, of course they are not. But this fixation on the afterlife has other—subtler but also pernicious—consequences.

The Islamic view of the relative insignificance of everything we see with our own eyes is that this world is merely a way station. While martyrdom is the extreme reaction, it is not the only reaction to this view of the world. The question arises: Why bother, if our sights are trained not on this life but on the afterlife? I believe that Islam's afterlife fixation tends to erode the intellectual and moral incentives that are essential for "making it" in the modern world.

As a translator for other Somalis who had arrived in Holland, I saw this phenomenon in various forms. One was simply the clash of cultures when immigrant Muslims and native-born Dutch lived in close proximity to one another. In apartment complexes, the Dutch were generally meticulous about keeping common spaces free of any litter. The immigrants, however, would throw down wrappers, empty Coca-Cola cans, and cigarette butts, or spit out the remnants of their chewed qat. The Dutch residents would grow incensed at this, just as they would grow incensed by the groups of children who would run about, wild and unsupervised, at all hours. It was easy for one family to have many children. (If a man can marry up to four wives and have multiple children with each of them, the numbers grow quickly.) The Dutch would shake their heads, and in reply the veiled mothers would simply shrug their shoulders and say that it was "God's will." Trash on the ground became "God's will," children racing around in the dark became "God's will." Allah has willed it to be this way; it is there because Allah has willed it. And if Allah has willed it, Allah will provide. It is an unbreakable ring of circular logic.

There is a fatalism that creeps into one's worldview when this life is seen as transitory and the next is the only one that matters. Why pick up trash, why discipline your children, when none of those acts is stored up for any type of reward? Those are not the behaviors that mark good Muslims; they have nothing to do with praying or proselytizing.

This, too, helps explain the notorious underrepresentation

of Muslims as scientific and technological innovators. To be sure, the medieval Arabic world gave us its numerals and preserved classical knowledge that might otherwise have been lost when Rome was overrun by the barbarian tribes. In the ninth century, the Muslim rulers of Córdoba in Spain built a library large enough to house 600,000 books. Córdoba then had paved streets, streetlamps, and some three hundred public baths, at a time when London was little more than a collection of mud huts, lined with straw, where all manner of waste was thrown into the street and there was not a single light on the public thoroughfares.[23] Yet, as Albert Hourani points out, Western scientific discoveries from the Renaissance on produced "no echo" in the Islamic world. Copernicus, who in the early 1500s determined that the earth was not the center of the universe but rather revolved around the sun, did not appear in Ottoman writings until the late 1600s, and then only briefly.[24] There was no Muslim Industrial Revolution. Today, there is no Islamic equivalent of Silicon Valley. It simply is not convincing to blame this stagnation on Western imperialism; after all, the Islamic world had empires of its own, the Mughal as well as the Ottoman and Safavid. Though it is unfashionable to say so, Islam's fatalism is a more plausible explanation for the Muslim world's failure to innovate.

Significantly, the very word for innovation in Islamic texts, *bid'a*, refers to practices that are not mentioned in the Qur'an or the *sunnah*. One hadith translated into English declares that every novelty is an innovation, and every innovation takes one down a misguided path toward hell. Others warn against general innovations as things spread by Jewish and Christian influences and by all those who are ruled by misguided and dangerous passions. Those who innovate should be isolated and physically punished and their ideas should be condemned by the ulema.[25] It was precisely this mentality that killed off astronomical research in sixteenth-century Istanbul and ensured that the printing press did not reach the Ottoman Empire until more than two centuries after its spread throughout Europe.

Zakir Naik, an Indian-born and -trained doctor who has become a very popular imam, has argued that, while Muslim nations can welcome experts from the West to teach science and technology, when it comes to religion, it is Muslims who are "the experts."[26] Hence, no other religions can or should be preached in Muslim nations, because those religions are false. But look more closely at his point: Naik is implicitly acknowledging the success of the West in this world. All Muslim nations have to offer, he concedes, is a near-total expertise on the subject of the next world.

Reasons to Live

There must be an alternative. In some ways, the words of Prime Minister Golda Meir of Israel are even more true today than when she spoke them: "We will only have peace with the Arabs when they love their children more than they hate us." I would only substitute for the word "Arabs," "Medina Muslims." For while the phenomenon of murderous martyrdom was once a peculiar feature of the Israeli-Palestinian conflict, it has now spread throughout the Muslim world. This exaltation of the afterlife as a tenet of Islam is in desperate need of reform.

In the early fall of 2013, more than 120 Muslim scholars from around the world signed an open letter to the "fighters and followers" of Islamic State, denouncing them as "un-Islamic."[27] Their letter, originally written in classical Arabic, makes the point that it is forbidden in Islam to kill emissaries, ambassadors, and diplomats, as well as the innocent. It even says it is "permissible" in Islam to be loyal to one's country. But the letter does not question the overall concept of martyrdom or challenge the primacy of the afterlife. Predictably, it has had a very limited impact. There are no IS fighters laying down their arms as a result of it; no would-be Western jihadists have been persuaded by it to abandon the search for martyrdom in Syria.

We need to go much further. Until Islam stops fixating on the afterlife, until it is liberated from the seductive story of life after death, until it actively chooses life on earth and stops valuing death, Muslims themselves cannot get on with the business of living in *this* world.

Perhaps Islam can take a page from the Protestant Reformation in this respect. As we have seen, the sociologist Max Weber theorized that Protestantism, though still focused on the afterlife, fostered a more constructive engagement with the world with the doctrine of "election," whereby the "godly" were deemed to have been preselected to be saved in the afterlife. Simply put, certain Protestant sects tended to encourage the decidedly capitalistic virtues of diligence, frugality, hard work, and deferred gratification. According to Weber, the Protestant ethic gave rise to a distinctive and transformative "spirit of capitalism" in North America and northern Europe.

Might a similar process be possible within the Islamic world? Could there be a comparable "Muslim ethic"—one that might lead in time to a greater engagement with this world? Perhaps. There is no doubt that Islam has its own commercial tradition. Muhammad himself was a caravan trader. There are entire chapters of sharia devoted to things such as contracts and rules for trade. And, as Timur Kuran has shown, sharia is not overtly hostile to economic progress; in the Ottoman Empire it established commerce-friendly legal rules and institutions. It was just that European legal systems were more conducive to capital formation.[28]

Explanations abound for the relative economic backwardness of many Muslim countries, ranging from corrupt governance to the "resource curse" of plentiful oil. But I am not one of those who think Muslims are condemned to economic failure. On the contrary, in countries such as Indonesia and Malaysia, there is ample evidence that a capitalist ethic can coexist with Islam. Anyone who takes the time to walk through a North African souk will see how readily Muslims engage in trade. As Hernando de Soto has noted, it was frustrated entrepreneurs, driven to self-immolation by the depredations of corrupt dictatorships, who launched the Arab Spring.

If imams started talking about making *this* world a paradise, rather than preaching that the only life that matters is the one that begins at death, we might begin to see economic dynamism in more Muslim-majority economies. Giving capitalism a greater chance to thrive in Islamic societies might be the most effective means of redirecting the aspirations of young Muslims to the rewards of life on earth instead of the promise of rewards after death. Such opportunities would give them a reason to live, instead of a reason to die. Only when Islam chooses *this life* can it finally begin to adapt to the modern world.

CHAPTER 5

SHACKLED BY SHARIA

How Islam's Harsh Religious Code Keeps Muslims Stuck in the Seventh Century

In Sudan a twenty-seven-year-old woman, Meriam Ibrahim, who was at the time eight months pregnant, was sentenced to suffer one hundred lashes and death by hanging for the crimes of adultery and apostasy. This sentence was not passed in 714 or 1414. It happened in 2014.

Meriam's crimes and my own are essentially the same under sharia. We both have been accused of leaving our religion. Like her, I married an infidel. I left religion entirely, whereas Meriam chose to follow the faith of her mother, an Ethiopian Christian, rather than her father, a Sudanese Muslim, and married a Christian man. Her "outing" by her family was an act of "commanding right and forbidding wrong," a practice with which we will deal in the following chapter, but her treatment after her arrest was determined in accordance with sharia. One of Meriam's own brothers told CNN that her husband had given her "potions" to convert her to Christianity and that, if she did not renounce her faith and repent, "she should be executed."[1]

Under Sudan's Islamic law code, and sharia in general, a father's religion is automatically the religion of his children. And Muslim women are prohibited from marrying outside their faith, although that prohibition does not apply to Muslim men. Thus, to the Sudanese sharia court, it did not matter that Meriam Ibrahim was raised as an Orthodox Christian by her mother. It did not matter that her father was absent for most of her childhood. It did not matter that she was married to an American citizen. In the strict application of Islamic law, apostasy is punishable by death, while adultery is punishable by one hundred lashes.

The sentence was not inflicted immediately because Meriam was pregnant when she was jailed—she gave birth to her daughter while shackled in leg irons to a wall in her cell. Sharia defers the imposition of the death penalty on a pregnant mother until her baby is ready to be weaned. Her only recourse, according to the Sudanese court, was to renounce Christianity and return to Islam. Indeed, in recent years, recanting and returning to Islam is how other apostates have avoided such a death sentence. But Meriam refused. Clerics were brought to visit her in jail, and she said she would not renounce Christianity for Islam. She said simply: "How can I return when I was never a Muslim?"

The U.S. State Department declared it was "deeply disturbed" by Meriam's harsh sentence. Condemnation also came from Amnesty International, and the embassies of Australia, Canada, the Netherlands, and the United Kingdom. It took months for the Sudanese government to grasp the scale of the public relations disaster it was inflicting on itself. Still the authorities sought to save face. Even after her death sentence was overturned, Meriam was accused of forging documents and was not allowed to leave Sudan. Instead, the "Agents of Fear," an element of the Sudanese police apparatus, trapped her at the airport. There, they beat up Meriam as well as her lawyers.

Only negotiations by Italian diplomats finally persuaded the Sudanese to relent, and Meriam's first stop after gaining her freedom was to meet with Pope Francis. (Here, incidentally, we see the stark difference between two faiths. In Argentina, the pope's birthplace, where Catholicism enjoys financial support from the state, are those who leave the Church sentenced to death? Are those who marry outside the Catholic faith convicted of adultery and sentenced to one hundred lashes?)

Abuses like those committed against Meriam are not isolated incidents. Sharia is routinely invoked or applied in all manner of circumstances across much of the Islamic world. And each time, its authority comes ultimately from Islam's sacred texts.

Here is a sampling of acceptable punishments under sharia:

Beheadings are sanctioned in chapter 47, verse 4, of the Qur'an, among others, which states, "when ye meet the Unbelievers (in fight), smite at their necks."

Crucifixions are sanctioned in 5:33: "The punishment of those who wage war against Allah and His Messenger, and strive with might and main for mischief through the land is: execution, or crucifixion, or the cutting off of hands and feet from opposite sides, or exile from the land."

Amputations are prescribed in 5:38: "As to the thief, Male or female, cut off his or her hands: a punishment by way of example, from Allah, for their crime: and Allah is Exalted in power."

Stonings are also permitted, according to the hadith Sunan Abu Dawud, book 38, no. 4413: "Narrated Abdullah ibn Abbas: The Prophet (peace be upon him) said to Ma'iz ibn Malik: Perhaps you kissed, or squeezed, or looked. He said: No. He then said: Did you have intercourse with her? He said: Yes. On the (reply) he (the Prophet) gave order that he should be stoned to death."

The Qur'an specifically urges Muslims not to be moved by compassion in cases of adultery and fornication, and decrees a public lashing. Chapter 24, verse 2, instructs: "The woman and the man guilty of adultery or fornication, flog each of them with a hundred stripes: Let not compassion move you in their case, in a matter prescribed by Allah, if ye believe in Allah and the Last Day: and let a party of the Believers witness their punishment."

Nor are beheadings, crucifixions, amputations, stonings, and lashings considered to be antiquated punishments. Some or all of them remain fully operational in countries such as Iran, Pakistan, Saudi Arabia, Somalia, and Sudan, where they are either sanctioned by the state or frequently imposed by the local faithful with tacit official approval. At the time of writing, the Saudi writer Raif Badawi is being subjected to the brutal punishment of public whipping because of blog posts judged blasphemous under sharia.

What Is Sharia?

Sharia formally codifies Islam's many rules. It governs not just how you worship, but also the organization of your daily life, your personal behavior, your economic and legal transactions, your life at home, and in many cases even the governance of your nation. The nineteenth-century French political theorist Alexis de Tocqueville, who was so astute in his understanding of American democracy, wrote: "Islam . . . has most completely confounded and intermixed the two powers . . . so that all the acts of civil and political life are regulated more or less by religious law."[2] Today, that same religious law remains the cornerstone of the Muslim world. It is exacting and punishment-centered. It prescribes what to do with unbelievers, both infidels and those who stray from the faith. It even contains rules on what types of blows are permissible when a husband beats his wife.

When we in the West think of the law, we conceive of it as a set of rules that govern the use of power and protect the rights of individuals. We have rules for everything, from driving to business contracts to the protection of private property, as well as rules to ensure fair treatment—to prevent individuals, corporations, and governments from acting recklessly, punitively, or without proper cause—and rules to punish those responsible for personal injury. The law evolves, a living thing that adapts to our changing society. The law also exists to resolve disputes. We settle in or out of court. But we settle peacefully.

Sharia arises out of an entirely different set of impulses. In early Islam, the state government was, as Patricia Crone describes it, "first and foremost about the maintenance of a moral order." The first allegiance in the Muslim community was to the imam, because only with a religious leader could the people "travel along the legal highways revealed by God." What separated Muslims from the infidels were not their laws; it was the God-given nature of their laws.[3] And because these laws came ultimately from Muhammad's divine revelations, they were fixed and could not be changed. Thus the law code dating from the seventh century continues to be followed today in nations and regions that adhere to sharia. Where Western laws generally set boundaries for what cannot be done, leaving everything else permissible, with sharia the system is reversed. The list of things that *can* be done is very small, while the list of what cannot be done overwhelms everything else—except for the list of punishments, which is even longer.

As a legal text, the Qur'an reflects its origins in a tribal or clan-based society, particularly on issues concerning inheritance, male guardianship, the validity of a woman's testimony in court, and polygamy. This is even more obvious in the hadith, the compilation of sayings attributed to the Prophet or documenting his actions. This combination of the Qur'an and the example of Muhammad forms the basis of sharia. The derivation of these legal rules, known as *fiqh*, is the responsibility of Islamic jurists and takes place on the basis of *ijma* (consensus). When conflicts of interpretation arise, scholars consult the Qur'an and hadith. If both are silent on the subject, jurists rely on a method of analogy (*qiyas*) to reach consensus.

As Ernest Gellner points out in his classic work, *Muslim Society*, "In traditional Islam, no distinction is made between lawyer and canon lawyer, and the roles of theologian and lawyer are conflated. Expertise on proper social arrangements, and on matters pertaining to God, are one and the same thing."[4] In other words, it is as if our priests, ministers, and rabbis were also our judges and legislators, employing their religious theology to establish legal boundaries of acceptable conduct in our daily lives.

Over the years, I have engaged in many discussions and debates on the Qur'an and hadith and their role in sharia. A common reply from devout Muslims is that the Bible (particularly the Old Testament book of Leviticus, but other sections as well) contains rules and punishments that are strict and stringent and antiquated by modern standards; thus it is unfair to single out Islam.

It is true that many parts of the Old and New Testaments reflect patriarchal norms. It is also true that the Hebrew scriptures contain many stories of harsh divine and human retribution. Even nonbelievers have heard of the concept of "an eye for an eye." In Deuteronomy, Moses imparts a great many laws, governing everything from the removal of boundary stones to the muzzling of oxen, to prohibitions against marrying one's stepmother, to the punishment of stoning for the crime of idolatry. The difference is that no one invokes these passages in modern-day jurisprudence, and their prescribed punishments have long since been set aside.

If there is one set of rules that is "timeless" in the Jewish Torah and the Christian Bible, it is the Ten Commandments, a relatively short list of prohibitions on killing, stealing, adultery, and so on. It was assumed that most laws were not religious in origin. Indeed, Judaism contains an ancient principle known as "*dina demalkhuta dina*," which means, "The law of the land is the law."[5] This is the principle that has made it possible for Jews as a community to exist under civil laws that differed from their own religious laws.[6] Christ, too, made it clear to his followers that they should "render up to Caesar what is Caesar's," and not only where Roman imperial taxation was concerned. Islam, on the other hand, views any law not in harmony with its own as illegitimate (5:44, 5:50). And its own law—sharia—derives from the entire Qur'an and all hadith.

Nothing drove this home to me more than the session of my Harvard seminar in which we discussed the drafting of a new Egyptian constitution. The Egyptian student who had previously shouted that she hadn't done the assigned reading declared: "It really doesn't matter what you write in the constitution of Egypt. It's not going to change anything. We are just going to carry on living the way we live."

Sadly, what she says is true. In Egypt, people rely on religious Muslim judges to decide contractual disputes and inheritance issues. When the military government wanted to condemn more than five hundred political prisoners to death—many of them members of the Muslim Brotherhood—it still needed a sharia court to sign off on the sentence.

At the other end of the spectrum are groups such as Boko Haram and IS, which believe they are reviving sharia as it was enforced by Muhammad and the first generation of his followers. When they stone, amputate, crucify, sell into slavery, or force religious conversions, they claim to be following the pure sharia code, and they can and do cite lines from it to justify their actions.

Global Sharia

"I did not kill! I did not kill!" a woman shrieks as Saudi police wrap her head with a black scarf.

"Praise God," a Saudi executioner dressed in white tells her.

He lifts his long silver sword and strikes her neck—a gasp, then she falls silent.

Twice more the hangman hacks at her neck, before stepping away to carefully wipe the blade.

Ambulance workers immediately start placing the woman's remains on a stretcher as charges against her are hurriedly read out over a loudspeaker in the Muslim holy city of Mecca.

> She was accused of raping her seven-year-old stepdaughter with a broomstick and beating her to death. "A royal decree was issued to carry out the sharia law, in accordance with what is right," the statement says.[7]

It is a remarkable fact that, after Friday prayers in Saudi Arabia, many men flock to the central squares to watch the implementation of Islamic justice: the cutting off of robbers' hands, the stoning of adulterers, and the beheading of murderers, apostates, and other convicted criminals.

Can anyone today imagine a congregation of Catholics leaving mass or Baptists leaving church or Jews leaving synagogue to go and spectate at a lethal injection or an electrocution? Though the death penalty is still inflicted in some U.S. states, we in the West have come very far since the days when public executions were the norm and religious offenses were punishable by death. Far from diminishing, this legal divide between Islam and the West is growing wider and deeper, and is increasingly global in scope.

When eighteen Palestinians in Gaza were shot dead in the summer of 2014 for allegedly collaborating with Israel, the immediate justification proffered was that these men had been found guilty by "local courts, supported by religious clerics." (Palestinian law makes collaborating with Israel a crime punishable by death, although the Palestinian president must give his approval before the sentence is carried out.) In other words, they had been tried and convicted under some version of the sharia system. And while human rights activists protested against the killings, there was no challenge to the underlying religious justification, or the role of Muslim clerics in approving these sentences.

In Pakistan, blasphemy against the Prophet Muhammad

is punishable by death.[8] More than thirty countries around the world have similar antiblasphemy laws, including some Christian ones. But it is in Muslim countries that such laws are enforced. In 2014, a Pakistani court sentenced a twenty-six-year-old Christian man to death on the ground that he had spoken ill of the Prophet. He argued that he was merely the target of fabricated accusations made by disgruntled local businessmen seeking to build an industrial center in his neighborhood. When his sentence was handed down, thirty-three other Pakistanis were already on death row for the crime of blasphemy.

What is more, with or without a formal court verdict, vigilantes are happy to carry out their own sentences. Again in Pakistan, a lawyer defending a university professor against blasphemy charges was shot and killed; in the southern city of Bahawalpur, after police had taken an alleged blasphemer into custody, a mob broke into the station, dragged the man into the street, and burned him alive as the law enforcement officers watched.

Such atrocities also occur in the supposedly more moderate nations of Southeast Asia. In the Indonesian province of Aceh, when a twenty-five-year-old widow was found with a forty-year-old married man, a group of eight local males beat the man, gang-raped the woman, and doused them both with sewage, before turning them over to the local sharia authorities. Then the sharia police handed down their own sentence: public caning for both the man and the woman for the crime of alleged adultery. In some ways, they got off relatively lightly; previously, the punishment for their crime had been death by stoning.[9]

It is of course tempting for the Western reader to assume that these are antiquated practices that, like witch-burning in Massachusetts, will eventually die out. But the trend in the Muslim world is in the other direction. In supposedly advanced Brunei, the ruling sultan is currently phasing in an "updated" body of sharia criminal law, making adultery punishable by stoning, theft punishable by amputation, and homosexual intercourse punishable by death. In Malaysia, which recognizes Islam as its official religion, supporters of Islamic law want to introduce sharia punishments, such as amputation for stealing, into the nation's penal code.

The modern trend for Islamic states to adopt more hard-line legal codes began with the formation of the Saudi kingdom, but it accelerated after the 1979 Iranian Revolution. In its aftermath, Iran became the first fully fledged modern Islamic theocracy. Iran's adoption of a strict Islamic legal code was popular at the time because it represented a clear and fundamental contrast to everything Iranians had abhorred about the regime of the deposed shah: its decadence, its corruption, its immorality.

Today, sharia has spread to a point where it has found near-universal acceptance across the Muslim world. Perhaps the most compelling evidence comes from the Pew Research Forum's 2013 report, "The World's Muslims: Religion, Politics, and Society," a study of thirty-nine countries and territories on three continents—Africa, Asia, and Europe—with more than 38,000 face-to-face interviews in eighty-plus languages and dialects, covering every country with more than 10 million Muslims. In response to the question "Do you favor or oppose making sharia law, or Islamic law, the official law of the land in our country?" the nations with the five largest Muslim populations—Indonesia (204 million), Pakistan (178 million), Bangladesh (149 million), Egypt (80 million), and Nigeria (76 million)—showed overwhelming support for sharia. To be precise, 72 percent of Indonesian Muslims, 84 percent of Pakistani Muslims, 82 percent of Bangladeshi Muslims, 74 percent of Egyptian Muslims, and 71 percent of Nigerian Muslims supported making sharia the state law of their respective societies. In two Islamic nations that are considered to be transitioning to democracy, the number of sharia supporters was even higher. Pew found that 91 percent of Iraqi Muslims and 99 percent of Afghan Muslims supported making sharia their country's official law.

Moreover, sharia is no longer restricted to Muslim-majority countries. It is increasingly being referenced in family law and inheritance cases involving Muslims in the West. Several sharia courts are now operational in Britain.[10] According to sharia, rather than inheriting equally under British common law, Muslim women can inherit only half what men inherit; divorced Muslim women cannot inherit at all, nor can adopted children, and non-Muslim marriages are not recognized.[11] Pressure to apply sharia in other Western nations is also on the rise. France, for example, has faced pressure over its laws forbidding polygamy when men from Muslim nations have wanted their second or third wives to immigrate to join them. So far the French authorities have refused to countenance polygamous marriages—just as France has also taken a stand against allowing Muslim girls to veil themselves at school. Still, some exceptions have already been made for children from polygamous marriages.

Support for sharia is also rising among Muslims living in the West. A 2008 survey of more than nine thousand European Muslims by the Science Center Berlin reported strong belief in a return to traditional Islam. In the words of the study's author, Ruud Koopmans, "Almost 60 per cent agree that Muslims should return to the roots of Islam, 75 per cent think there is only one interpretation of the Quran possible to which every Muslim should stick and 65 per cent say that religious rules are more important to them than the laws of the country in which they live."[12] More than half (54 percent) of those surveyed also believe that the West is out to destroy Muslim culture.[13]

The Sharia Paradox

One of the Saudi Arabian kingdom's leading executioners, Muhammad Saad al-Beshi, told the publication *Arab News* that he has executed as many as ten people in a single day. The sword is his preferred instrument. He keeps his blade "razor sharp" and has his children help him keep it clean. Al-Beshi finds it interesting that people are amazed at how fast the sword can separate the head from the body, and he wonders why people come to watch executions if they are going to faint and "don't have the stomach for it." Al-Beshi also carries out the rulings of sharia by severing hands, feet, and tongues.

Such comments must be deeply shocking to most Western readers, even those living in countries that retain some form of capital punishment. Yet for years, like most Muslims, I myself did not think to question the basic principles and practices of sharia. Even in running away from my arranged marriage, I believed that sharia punishments would follow me because that was the rule in my own community. When I arrived in the Netherlands, I feared that my father or his clansmen or the man I had been assigned to marry would simply appear and force me to submit against my will. When the Dutch officials first told me that there were laws to protect me and that the Netherlands would not recognize my arranged marriage because it had no legal standing, I marveled at this system, so different from the Islamic code. As I immersed myself more in the beliefs and teachings of Western liberal thought, I only marveled more.

In small seminars at Leiden, we reflected on World War II. Did ordinary Germans know about the Holocaust? Did the Dutch? We were put in the position of asking ourselves: What would I have done in the circumstances? Would I have been a "willing executioner"? Would I have helped the Jews, at the risk of my own life? Would I simply have done nothing? As I was grappling with these agonizing questions, my younger sister—who had joined me in the Netherlands—was going through what I had experienced in Nairobi. It was her turn now to feel that she must strive to be a good and pious Muslim. She was reading Sayyid Qutb's *Milestones* and Yusuf al-Qaradawi's *The Lawful and Prohibited in Islam*: key texts of the Muslim Brotherhood. She was embracing sharia, even as I was being taught to understand the importance of man-made laws and the appalling consequences of lawless totalitarianism.

Today, thanks in large part to my years at Leiden, I understand—I *know*—that each person, regardless of sex, orientation, color, or creed, deserves basic human rights and protections in return for adhering to the laws of the land where they live. But I also know that this truth contradicts many of the fundamental dictates of sharia. Whereas the rule of law in the West evolved to protect the most vulnerable members of society, under sharia it is precisely the most vulnerable who are also the most constrained: women, homosexuals, the insufficiently faithful or lapsed Muslims, as well as worshippers of other gods.

Consider the following crimes and their appropriate punishments as dictated by the Qur'an:

- Apostasy: the penalty for leaving the Islamic "tribe," is death. "If they turn renegades, seize them and slay them wherever ye find them" (4:89).
- Blasphemy: the Qur'an does not identify an exact punishment on earth, but notes in 9:74, "Allah will punish them with a grievous penalty in this life and in the Hereafter. They shall have none on earth to protect or help them." (See also 6:93 and Sahih al-Bukhari, volume 5, book 59, no. 369.)
- Homosexuality: according to the hadith, "If you find anyone doing as Lot's people did, kill the one who does it, and the one to whom it is done." (Sunan Abu Dawud, book 38, no. 4447.)

No group is more harmed by sharia than Muslim women, however—a reflection in part of the patriarchal tribal culture out of which Islamic law emerged. Repeatedly, women are considered under the code to be worth at most "half a man." Sharia subordinates women to men in a multitude of ways: the requirement of guardianship by men, the right of men to beat their wives, the right of men to have unfettered sexual access to their wives, the right of men to practice polygamy, and the restriction of women's legal rights in divorce cases, in estate law, in cases of rape, in court testimony, and in consent to marriage. Sharia even states that women are considered naked if any part of their body is showing except for their face and hands, while a man is considered naked only between his navel and his knees.[14]

Not untypical of the transgressions sharia identifies is that of the "rebellious wife," who is defined in the seminal Sunni commentary *Reliance of the Traveller: A Classic Manual of Islamic Sacred Law* as a woman who merely answers her husband "coldly, when she used to do so politely." The husband, the book states, should start to reprimand her with the verbal warning "Fear Allah, concerning the rights you owe to me." If that fails, he can stop speaking to her and then may strike her, although not to "break bones, wound her, or cause blood to flow."[15]

One of the most onerous burdens sharia imposes on women is the principle of guardianship. It is based on both a series of Qur'anic verses and the commentaries in the accompanying hadith. In essence, guardianship is presented as a way to protect women, but in reality it forces women to be entirely dependent on male guardians for the most basic activities outside the home, from shopping for their families to visiting the doctor. The Qur'an states, in 4.34: "Men are the maintainers [guardians] of women. . . . As to those women on whose part ye fear disloyalty and ill-conduct, admonish them (first), (Next), refuse to share their beds, (And last) beat them (lightly); but if they return to obedience, seek not against them Means (of annoyance): For Allah is Most High, great (above you all)."

Chapter 2, verse 223 also categorizes women as "like a tilth for you," which is interpreted in sharia as ensuring that a husband shall have unfettered sexual access to his wife or wives, provided they are not menstruating or physically ill. Polygamy, too, is asymmetrical under sharia, as it is in all traditional patriarchal societies. Men, according to the Qur'an, may take up to four wives, but women may take no more than one husband.

A girl can be married off without her consent by her father or guardian. After she reaches puberty, seeking her permission is recommended but not required, and her silence is considered permission. According to *Reliance of the Traveller*, even if a woman has selected a "suitable match" for herself, she will be overruled if her guardian has chosen a different suitor who is also a suitable match.[16] In practice, many Muslim girls are married off long before they form a view of their own on the matter. In countries that adhere to a strict form of sharia, the marital age is often lowered, following in the tradition of Muhammad, who married his wife Aisha when she was six or seven years old and consummated that marriage when she was nine (she moved into Muhammad's house with her dolls, according to one of the hadith). Yemeni fathers, for example, routinely marry off their daughters by the age of nine, on the ground that this prevents adultery. Finally, although Muslim men may marry Christian or Jewish women, Muslim women may only marry Muslim men. And, as we have seen, the penalties for breaking this law can be harsh.

The inequality of the sexes, in short, is central to sharia. The Qur'an says that a son shall inherit as much as two daughters. In a sharia-based court, to prove the crime of rape, either the rapist must confess or four male witnesses must come forward to say that they each saw the rape take place. As a general rule, 2:282 of the Qur'an says that a woman's testimony is worth only half of a man's testimony in court. And, while men may easily divorce their wives under Islamic law—simply by saying the words "I divorce you!" three times—it is far more difficult for a woman to secure a divorce from her husband. Women also lose custody when a child turns seven, whereas men do not.

This is not a history book about past practices. These are contemporary laws and contemporary punishments, taking place in the twenty-first century. And I believe it is these practices—not young mothers like Meriam Ibrahim—that need to be condemned and shackled.

The Honor/Shame Dynamic in Sharia

Given Islam's origins among the clans and tribes of Arabia, we should not be surprised that there is also a strong emphasis on honor in sharia. In particular, the interaction of the principle of male guardianship with tribal norms of modesty frequently leads to "honor" violence against women (see chapter 6).[17]

It is true that honor violence is not an exclusively Muslim phenomenon. It is also true that honor killings predate Islam. Yet honor killings are common in the Muslim world and Islamic clerics have shown a tacit acceptance of them.[18] An honor killing is, in effect, a crime without a punishment according to *Reliance of the Traveller*, which explicitly exempts parents who kill their children from any accountability.[19] Such attitudes have proved remarkably durable. In 2003 the Jordanian Parliament voted against a bill that would have established harsher legal penalties for honor killings on the ground that it would violate "religious traditions." When a committee in the Senate then proposed to apply the same leniency shown to men who commit honor killings to women who kill husbands caught in adultery, the Muslim Brotherhood in Jordan strongly objected.

The arguments presented are worth noting for the connection they make between a woman's religious virtue and the bloodline. Sheikh 'Abd al-'Aziz al-Khayyat, a former Jordanian minister of religious affairs (*awqaf*), even issued a *fatwa* (Islamic religious ruling) stipulating that sharia does not give a wife the right to murder her husband if she catches him with another woman. Such a case, Khayyat explained, does not amount to an offense against the family's honor but only against the couple's marital life, and the most the wife is allowed to do is to file for divorce. Another Jordanian lawmaker, 'Abd al-Baqi Qammu, explained: "Whether we like it or not, women are not equal to men in Islam. Adulterous women are much worse than adulterous men, because women determine the lineage."[20]

Such open justifications of violence against women are remarkably easy to find. On Egyptian television during a 2010 talk show, a Muslim cleric, Sa'd Arafat, reviewed the rules for beating one's wife. He began by saying, "Allah honored wives by installing the punishment of beating."[21] Beating, he explained, was a legitimate punishment if a husband did not receive sexual satisfaction from his wife. But he added: "There is a beating etiquette." Beatings must avoid the face because they should not make a wife ugly. They must be done at chest level. He recommended using a short rod.

If that sounds almost comical it should not distract us from the shocking reality that violence against women has surged in Egypt since the Arab Spring. When supporters of President Abdel Fattah el-Sisi gathered in Cairo's Tahrir Square to celebrate his inauguration in June 2014, dozens of women were sexually assaulted, and a nineteen-year-old was brutally gang-raped. These crimes were incited by Islamist preachers such as the Salafi Abu Islam, who said that any women going uncovered to Tahrir Square "want[ed] to be raped."

Nor is it only women who are discriminated against under sharia. More than thirty Islamic countries have state laws that prohibit homosexuality and make it a criminal offense, punishable by everything from lashing to life imprisonment. In Mauritania, Bangladesh, Yemen, parts of Nigeria and Sudan, the United Arab Emirates, Saudi Arabia, and Iran, convicted homosexuals can be sentenced to death. In Saudi Arabia, a man found guilty of homosexual activity may be executed or he may receive a hundred lashes and a lengthy prison sentence. In Iran, men who play "an active role" receive a hundred lashes, while the "recipient" can be put to death. For lesbians, the punishment is one hundred lashes; after four convictions, it is death.[22] A 2012 study by an Iranian human rights group (IRQO) in association with the International Human Rights Clinic at Harvard Law School found that some lesbian, gay, bisexual, and transgender individuals in Iran are openly forced to undergo gender reassignment surgery.[23]

Death by Stoning

Sharia also sanctions the odious punishment of stoning, a practice that should be unthinkable in this century, yet remains far too common. Today at least fifteen countries and territories have laws that allow or require death by stoning, particularly for crimes of adultery or other forms of "sexual promiscuity." A survey for the Pew Institute in 2008 found that only 5 percent of Pakistanis opposed stoning for adultery; 86 percent were in favor of it.[24]

Iran has the highest per-capita rate of stonings in the world. Under its legal system, judges are allowed to convict a defendant based not on evidence but on a "gut feeling" of guilt. In an odd echo of the religious persecutions of the European Middle Ages, when the accused could prove his or her innocence only by surviving an ordeal such as walking over burning stones or being immersed in frigid water, present-day Iranian stoning victims can survive only if they can escape. But whereas men are buried up to their waists, making escape an option for the strong and swift, women are usually buried up to their chests, wearing their chadors, making escape all but impossible.

Stoning occurs all over the Muslim world. In Tunisia, the Commission for the Promotion of Virtue and the Prevention of Vice demanded the stoning of a nineteen-year-old who had posted nude images of herself online. In my homeland of Somalia, a thirteen-year-old girl reported that she had been gang-raped by three men. The Al-Shabaab militia that then controlled her town of Kismayo, a port city in the south, responded by accusing *her* of adultery, found her guilty, and sentenced her to death. Her execution was announced in the morning from a loudspeaker blaring from a Toyota pickup truck. At the local soccer stadium, Al-Shabaab loyalists dug a hole in the ground and brought in a truckload of rocks. A crowd of one thousand gathered in the hours leading up to 4:00 p.m. Aisha Ibrahim Duhulow—named after the Prophet Muhammad's nine-year-old wife—was dragged, screaming and flailing, into the stadium.[25] It took four men to bury her up to her neck in the hole. Then fifty men spent ten minutes pelting her with rocks and stones. After the ten minutes had passed, there was a pause. She was dug out of the ground and two nurses examined her to see if she was still alive. Someone found a pulse and breathing. Aisha was returned to the hole and the stoning continued. One man who tried to intervene was shot; an eight-year-old boy was also killed by the militia. Afterward, a local sheik told a radio station that Aisha had provided evidence, confirmed her guilt, and "was happy with the punishment under Islamic law."

In 2014, a group called Women Living Under Muslim Laws circulated a petition to the United Nations, asking that body to enact international laws against stoning. They collected a paltry 12,000 signatures. While some Muslim clerics disavow stoning, others say the hadith supports it, while still others argue that Muhammad was merely following contemporary Jewish practice. These arguments are all presented as rational positions, as if there is a debate worth having on the subject. But how can there be any position on stoning other than that it is barbaric and evil?

The classic Western response to relativist arguments was offered by Sir Charles Napier, who in 1842 was appointed commander of British forces in India. When local religious authorities complained against the banning of *sati*, explaining that it was the Hindu custom to burn alive the wife of a man who had died, Napier replied: "My nation also has a custom. When men burn women alive we hang them, and confiscate all their property. . . . Let us all act according to national customs." Today, however, such an exchange is scarcely imaginable. Instead, Western authorities bend over backward to accommodate Muslim "sensitivities" and often excuse or look the other way when Muslims violate universal human rights—even when they do so in our own countries.

Needed: A New Language of Emancipation

Beyond the ways it restricts women's rights and legitimizes violence against them, sharia does something more. Because of the very foundation of sharia in the dictates of the Qur'an and the hadith, there is no vocabulary in Islam that can be used to emancipate women. All the words for female rights and basic female freedoms are invariably Western. If you fight for access to education or the right to vote or the right to drive or the right not to be beaten or stoned, the vocabulary you have to use in making that argument is Western because Islamic texts and the Arabic language simply do not have the words for these types of rights and opportunities. By contrast, when women face opposition to their emancipation, those words and that vocabulary are exclusively Islamic. In Somalia, people say to women who do not want to be in polygamous marriages, "Oh, yeah, sure, you want to be just like the *gaalo*." The *gaalo* are the infidels, a derogatory term that means being unfaithful to God. So if you don't want to be a second or third wife, or you don't want to be replaced by a second or third wife, you are simply being unfaithful to God. It is almost impossible to have a discussion about these issues that doesn't bring Islam into the conversation. People say, "It's ungodly, it's not what the Prophet Muhammad said to do."

This is not to say that women have a long history of being fully emancipated in the West. Until well into the 1970s, as is well known, a married woman couldn't even open a charge card at a Sears store in her own name. Historically, some of the most vocal forces opposing the emancipation of American women came from the Christian clergy. Many argued that the subservience of women was a God-given fact, and that to release women from the home would lead to the enslavement of men. Yet there were equally convinced clergymen on the other side. Reverend Theodore Parker of Boston said in 1853, "To make one half of the human race consume its energies in the functions of housekeeper, wife, and mother is a monstrous waste of the most precious material God ever made."[26] In Islam, by contrast, such arguments are scarcely ever heard.

Cultural relativists prefer to wrap the issue of sharia in the intellectual equivalent of a black *jilbab* or blue burqa and intone the old platitudes that we should be nonjudgmental about the religious practices of others. Why? The ancient

Aztecs and other peoples practiced human sacrifice, tearing the still-beating hearts out of their sacrificial victims. We teach our children that this happened five hundred years ago, but we don't condone it—and wouldn't if the practice were suddenly revived in Mexico today. So why do we condone the "sacrifice" of women or homosexuals or lapsed Muslims for "crimes" such as apostasy, adultery, blasphemy, marrying outside of their faith, or simply wishing to marry the partner of their choice? Why, aside from the publication of reports by human rights organizations, is there no discernible reaction?

In the twenty-first century, I believe that all decent human beings can agree that such barbarous acts should not be tolerated. They can and must be condemned and prosecuted as crimes, not accepted as legitimate punishments.

The abuses carried out under sharia are irrefutable. If we are to have any hope for a more peaceful, more stable planet, these punishments must be set aside.

There is probably no realistic chance that Muslims in countries such as Pakistan will agree to dispense with sharia. However, we in the West must insist that Muslims living in our societies abide by our rule of law. We must demand that Muslim citizens abjure sharia practices and punishments that conflict with fundamental human rights and Western legal codes. Moreover, under no circumstances should Western countries allow Muslims to form self-governing enclaves in which women and other supposedly second-class citizens can be treated in ways that belong in the seventh century.

Yet that is not enough. We must also address and reform Islam's most powerful social tool: the informal grassroots enforcement of its strictest religious principles in the name of commanding right and forbidding wrong.

CHAPTER 6

SOCIAL CONTROL BEGINS AT HOME

How the Injunction to Command Right and Forbid Wrong Keeps Muslims in Line

When I was a teenage girl growing up in Nairobi, I wondered aloud in our house why the ritual prayers had to be said five times a day. Why not cut the number down to once a day? My half sister overheard me talking and almost immediately launched into hours of lectures, not just on that day but on many subsequent days, about my failures to perform my sacred duty as a Muslim. Nor did she confine herself to lecturing me. She also went about lobbying my extended family to have me "sent away" to be treated for "madness" because I had dared to ask a question about our faith and its practice.

This illustrates how the practice of commanding right and forbidding wrong functions in Islamic society. Debate and doubt are intolerable, deserving of censure, with the questioner reduced to silence even inside her own home. My half sister believed it to be her duty and obligation to correct me: to command me to do right and forbid me to do—or even think—wrong.

This is only part of a larger truth about Islam. It is almost always the immediate family that starts the persecution of freethinkers, of those who would ask questions or propose something new. Commanding right and forbidding wrong begins at home. From there, it moves out into the community at large. The totalitarian regimes of the twentieth century had to work quite hard to persuade family members to denounce one another to the authorities. The power of the Muslim system is that the authorities do not need to be involved. Social control begins at home.

The constant personal and intellectual unease that many of the Muslim students in my Harvard seminar felt with any discussion of the political organization of the Islamic world is directly connected to this overarching concept of commanding right and forbidding wrong. When the Qatari man challenged me on the first day of class, he was following these principles. He was not the last to do so. I had a male student from Nigeria who claimed to be an expert in sharia, among other things. He, too, repeatedly rose to "correct" me, each time calling me "sister," to emphasize the kinship element—although I was no doubt an apostate to him—and thereby also attempt subtly to nullify my role as the seminar leader. Women and men have very specified roles in Islamic society. It is spelled out exactly how each sex should act. And a man has an unequivocal right to command a woman, even if that woman is purportedly his teacher.

In short, taken together, commanding right and forbidding

wrong are very effective means of silencing dissent. They act as a grassroots system of religious vigilantism. And their most zealous enforcers find in these words an excuse not just to command and to forbid but also to threaten, to beat, and to kill. I think of it as the totalitarianism of the hearth.

Origins of Commanding Right and Forbidding Wrong

As far back as the philosophy of Aristotle and the Stoics in ancient Greece, Western civilization has understood the concept that the law must "command what should be done and forbid what should not be done." Thus the underlying concept of commanding right and forbidding wrong is not completely unique to Islam. The historian Michael Cook even speculates that "this ancient wording, like the owl on Athenian coins, found its way to pre-Islamic Arabia" from ancient Greece.[1]

Whatever the origin of the phrase, however, Muhammad's interpretation of it is explicit and novel. The Qur'an itself spells out the concept in three different places: "Let there arise out of you a band of people inviting to all that is good, enjoining what is right, and forbidding what is wrong: They are the ones to attain felicity" (3:104). "Ye are the best of peoples, evolved for mankind, enjoining what is right, forbidding what is wrong, and believing in Allah" (3:110). And later: "The Believers, men and women, are protectors one of another: they enjoin what is just, and forbid what is evil" (9:71).

Some scholars have argued that these Qur'anic definitions might mean little more than separating believers in Islam from nonbelievers, "right" entailing choosing the faith of Allah and "wrong" the decision to worship anything else. But that is not how the injunction has usually been interpreted.

Of course, all religions have rules. Some Protestant sects were especially intrusive in policing their members, as the early history of New England confirms. But the comprehensive nature of commanding right and forbidding wrong is uniquely Islamic. And because Islam does not confine itself to a separate religious sphere, it is deeply embedded in political, economic, and personal as well as religious life. As Patricia Crone explains, "Islamic law obliged its adherents to intervene when they saw other believers engage in sinful behavior and to persuade them to stop, or even to force them to do so if they could." The importance of this function was even comparable with that of jihad, because for the Muslims of that era, "fighting sinners and fighting infidels were much the same." In its practical application during the medieval era, commanding right and forbidding wrong entailed the Islamic ruler hiring a censor and market inspector who "would patrol the streets with armed assistants to ensure that people obeyed the law in public," whether it was attending Friday prayers, fasting during Ramadan, maintaining modesty in dress, forgoing wine, or segregating men and women.[2]

Remarkably, more than a thousand years later, little has changed. The religious police in Iran and Saudi Arabia, who beat women for displaying an ankle in public, the followers of the British-born lawyer and imam Anjem Choudary who carry out vigilante Muslim patrols in London,[3] chastising women for refusing to cover up and knocking alcohol out of adults' hands, and the sharia brigades cracking down on alcohol consumption in Wuppertal, Germany,[4] are the twenty-first-century commanders of right and forbidders of wrong. Today, as much as in medieval times, the concepts of commanding right and forbidding wrong entail telling individual Muslims how to live, down to the most intimate aspects of their lives.

Commanding Right and Forbidding Wrong in Practice

At its most extreme, the concept of commanding right and forbidding wrong provides the justification for fathers, brothers, uncles, and cousins who carry out honor killings of female relatives they believe have committed irredeemable transgressions. In many parts of the Islamic world, any behavior deemed immodest is reason enough to kill a daughter or female relative. And immodesty is extremely broadly defined: it could include singing, looking out a window, or speaking to a man who is not a relative. Marrying for love, in defiance of one's parents, is also a frequent justification.

No one knows the exact number of honor killings that happen around the world every year. Five thousand is the most commonly cited estimate, but that number illustrates only that the practice is underreported. The practice has certainly become more prevalent since the late twentieth century as more and more nations have formally adopted sharia. Almost a thousand honor killings occur annually in Pakistan alone.[5] The problem is that honor killings are often not reported, or are ignored, or are disguised. There is often little or no incentive to bring them to the authorities in countries where the authorities sanction them.

What does honor violence look like in practice? In Lahore, Pakistan, a twenty-five-year-old woman who married against her father's wishes was stoned to death outside a courthouse.

Also in Pakistan, a girl was shot dead while doing her homework because her brother had thought she was with a man. A Pakistani father and mother doused their fifteen-year-old daughter with acid because she had looked twice at a boy who passed by on a motorcycle, and from that they "feared dishonor." Her mother said that her daughter cried out before she died, "I didn't do it on purpose. I won't look again."[6] But the mother added, "I had already thrown the acid. It was her destiny to die this way." When seventeen-year-old Rand Abdel-Qader's father killed her in Basra, Iraq, because she had allegedly fallen in love with a British soldier stationed there, local officials commented: "Not much can be done when we have an honor killing case. You are in a Muslim society and women should live under religious laws."[7]

Farzana Parveen was three months pregnant when she was stoned to death in Pakistan in 2014 by her father, brother, and a family-selected fiancé whom she had declined to marry. Farzana had married against her family's wishes, the family felt shamed, so they killed her in broad daylight outside a courthouse in the city of Lahore. Even more appalling, she was the second woman to die in this case. Her husband had strangled his first wife so that he could marry Farzana. He paid blood money, it was deemed an honor killing, and so he was free to wed again. When Farzana was killed, her stoning was also deemed an honor killing.

A young mother of two in Punjab province was stoned to death by her uncle and cousins, using stones and bricks, on the order of a Pakistani tribal court simply because she had a cell phone. Even though stoning is supposedly illegal in Afghanistan, 115 men stood and cheered the stoning of a twenty-one-year-old woman accused of "moral crimes."

Commanding right and forbidding wrong can also justify the murder of homosexuals and Muslim apostates—even Muslims who are insufficiently devout. When the governor of Punjab acted to protect a Christian woman who was charged with blasphemy, it was his own bodyguard who killed him. Afterward, thousands of Pakistanis, including numerous clerics, lauded the killer, showering him with petals and celebrating his steadfastness and courage. Dawood Azami of the BBC's World Service explains the dangers of apostasy in Afghanistan:

> For those who were born Muslim, it might be possible to live in Afghan society if one does not practice Islam or even becomes an "apostate" or a "convert." They are most probably safe as long as they keep quiet about it. The danger comes when it is made public that a Muslim has stopped believing in the principles of Islam. There is no compassion for Muslims who "betray their faith" by converting to other religions or who simply stop believing in one God and the Prophet Muhammad. Conversion, or apostasy, is also a crime under Afghanistan's Islamic law and is punishable by death. In some instances, people may even take matters into their own hands and beat an apostate to death without the case going to court.[8]

Yet while these are striking examples, the practice of commanding right and forbidding wrong is subtler and more pervasive than they imply. In a 2013 profile of King Abdullah of Jordan, the writer Jeffrey Goldberg recounted a visit he made with the king to the Jordanian city of Karak (Abdullah flew his own Black Hawk helicopter), "one of the poorer cities in a distressingly poor country." The king was going to have lunch with the leaders of Jordan's largest tribes, which in Goldberg's words "form the spine of Jordan's military and political elite." It is a long-standing symbiotic alliance between the Hashemite kings and their kingdom's clan chiefs. The tribal leaders expect the king to help safeguard their power and privileges, in part by keeping Jordan's Palestinian population in check. In return, the tribes help to safeguard the king.

This particular trip was designed in part for Abdullah to make his pitch for developing viable political parties among the tribes before upcoming parliamentary elections. Having watched the chaos engulfing his neighboring nations and having seen the bloody overthrow of established (albeit nonroyal) rulers in Egypt, Libya, and Tunisia, Abdullah was hoping to mobilize the tribal leaders to stem the rise of the Muslim Brotherhood in Jordan and prevent it from "hijack[ing] the cause of democratic reform in the name of Islam." Still, his expectations were not high. Goldberg quotes the king as saying: "I'm sitting with the old dinosaurs today."

The meal was a traditional Bedouin one, eaten with forks (a small concession to modernity) at a long, high communal table, a hallmark of tradition. Then, with the ceremonial lunch complete, it was time for the tea and talk. Goldberg writes:

> The king made a short plea for economic reform and for expanding political participation, and then the floor was opened. Leader after leader—many of whom were extremely old, many of whom merely had the appearance of being old—made small-bore requests and complaints. One of the men proposed an idea for the king's consideration: "In the old days, we had night watchmen in the towns. They would be given sticks. The government

should bring this back. It would be for security, and it would create more jobs for the young men."[9]

"I was seated directly across the room from the king," Goldberg adds, "and I caught his attention for a moment; he gave me a brief, wide-eyed look. He was interested in high-tech innovation, and in girls' education, and in trimming the overstuffed government payroll. A jobs plan focused on men with sticks was not his idea of effective economic reform. As we were leaving Karak a little while later, I asked him about the men-with-sticks idea. 'There's a lot of work to do,' he said, with fatigue in his voice."[10]

But here's the rub: employing men with sticks is not some quaint old idea; it is a central component of Islam. Commanding right and forbidding wrong is in many ways all about men wielding sticks, enforcing correct behavior.

The Zone of Privacy Is Now a Dead Zone

Part of what makes commanding right and forbidding wrong such a menace is that, unlike the term "jihad," it sounds so virtuous. What could be wrong with living a moral life? Isn't that the primary aspiration of all major religious teachings? And what could be more reasonable than a devolved discipline, with norms of behavior enforced by family rather than some external power?

The problem is that these questions expose some fundamental differences between Islam and Western liberal thought. A core part of the Western tradition is that individuals should, within certain limits, decide for themselves what to believe and how to live. Islam envisages the exact opposite: it has very clear and restrictive rules about how one should live and it expects all Muslims to enforce these rules. In its modern conception, commanding right has become (in the words of Michael Cook) "the organized propagation of Islamic values."[11] As Dawood Azami puts it, if you depart from the basic (and time-consuming) requirements of the faith, you had best "keep quiet about it" if you hope to survive unscathed even by your own family.

It was not always this way. In the medieval period, there were disagreements about how far commanding and forbidding should extend. Behind closed doors, in private lives, without witnesses, there was more latitude. As Patricia Crone notes, "Freethinkers could discuss their views with like-minded individuals in private salons, in learned gatherings at the court, and to some extent in books and even more so in poetry, where things could be put ambivalently." There was even an entire Islamic literary style, the *mujun*, which allowed its practitioners to push the boundaries of what was acceptable in society, allowing them to teeter on the edge of the blasphemous, the pornographic, the scurrilous. "In short," Crone concludes, "freedom lay essentially in privacy. The public sphere was where public norms had to be maintained, where there might be censors or private persons fulfilling the duty of 'commanding right and forbidding wrong' who would break musical instruments, pour out wine, and separate couples who were neither married nor closely related. But their right to intrude into private homes was strictly limited."[12] There was even a way to say, to those who sought to enforce the Qur'an's dictates, "Mind your own business."

The idea of a zone of privacy and the concept of "mind your own business" have eroded in our time. As modern Islamic communities have become radicalized, there is a kind of arms race of commanding right and forbidding wrong. This means that a closet atheist is quickly outed because he is soon caught not praying five times a day, not fasting in the month of Ramadan, not praising Allah constantly, not saying "Inshallah" every time he refers to the future. While we in the West have surrendered our privacy to our credit card companies, website cookies, social media networks, and search engines, in the Muslim world the zone of privacy has been eroded by other means.

How Does This Doctrine Take Root?

Universal human rights also play no part in the conception of commanding right and forbidding wrong; there are only the rules of Islam. This phenomenon is at its most extreme with the so-called Islamic State, which demands that anyone living within its "caliphate" convert to its extreme practice of Islam and follow its rules. When IS fighters rolled into the city of Mosul, hanging out of car windows or off the backs of trucks, video footage captured one fighter aggressively wagging his finger at a woman on the street. He was signaling to her to cover up. Next would come the order for women not simply to cover, but to stay in their homes. Clothing stores in captured cities and towns could no longer sell anything but Islamic dress and all mannequins were to be veiled and covered.

How can formerly progressive cities and regions, or at least fairly modern ones, allow the clock to be turned back to such an extreme degree? The answer is that the central elements of this type of fundamentalism are already present in Islamic politics, albeit in diluted form. The IS agenda is in some respects not so different from that of the Muslim Brotherhood or the Saudi Wahhabist teachings; it is just that their methods are more exposed.

A particularly unfortunate legacy of the U.S.-led invasion that ousted Saddam Hussein was the rise of sectarian political parties and militias in the wake of the collapse of the single-party Ba'athist authoritarian state. What is clear in hindsight is that the Ba'ath party had not eradicated these beliefs; it had merely driven them underground. Once freed and unleashed, these groups and their clerics proclaimed honor killings to be a legitimate religious means of "policing" women's behavior. Islamists in Basra scrawled graffiti that read, "Your makeup and your decision to forgo the headscarf will bring you death." Years before 2014, in other words, the fundamentalist seeds were already there.

Syria, too, was widely regarded in the West as relatively secular. But the secularization has melted in the heat of civil war. In Raqqa, the Syrian city that became IS's capital, the insurgents have tested a sort of "Taliban 2.0" style of female repression. As in other fundamentalist states, women who go out without a male chaperone, or who are not fully veiled, are arrested and beaten; but in Raqqa, these arrests and beatings are frequently committed by other women. IS has invented something new in the history of commanding right and forbidding wrong: an all-female moral police, the Al-Khansaa Brigade. The philosophy behind the brigade is simple, according to Abu Ahmad, an IS official in Raqqa, who said in an interview, "We have established the brigade to raise awareness of our religion among women, and to punish women who do not abide by the law. Jihad," he added, "is not a man-only duty. Women must do their part as well."[13]

For the modern-day jihadists, embracing the doctrine of commanding right and forbidding wrong also provides an opportunity to expand their ranks and incorporate more individuals outside of a purely combatant role. The practice creates many more soldiers for Allah and, in the case of Al-Khansaa, creates new ways to manage women who cannot go off to traditional war. (At least not yet—the Norwegian Islamic terror expert Thomas Hegghammer foresees a gradual shift to give women "more operative" roles in the jihad fight, explaining: "There is a process of female emancipation taking place in the jihadist movement, albeit a very limited (and morbid) one.")

A teenage girl in Raqqa described to the publication *Syria Deeply* how the female IS brigades function in practice. She was simply grabbed from the street by a group of armed women. "Nobody talked to me or told me the reason for my detention," she told the reporter. "One of the women in the brigade came over, pointing her firearm at me. She then tested my knowledge of prayer, fasting, and hijab." This girl's "crime" was walking without an escort and with an improperly worn headscarf.

When life is dominated by the fear of small infractions, how little thought can be given to the bigger questions? For want of a properly tied headscarf, a woman is beaten. It is the theological counterpart of the American policing theory of fixing broken windows and getting panhandlers off the streets as a way to prevent petty crimes from leading to larger, more serious violent transgressions. In the theory of commanding right and forbidding wrong, every small act, every minor infraction has the potential to become a major religious crime. Who can think about rights or education or economics when a trivial sartorial lapse can have such monumental consequences?

In Iraq, too, the current political tumult has created

opportunities for vigilantism dressed up as religious policing. The dangers for gay Iraqi men are far greater today than they were under Saddam Hussein's regime. As *The Economist* notes, "Men even suspected of being gay face kidnappings, rape, torture and extrajudicial killing" by self-appointed sharia judges and squads that deem themselves to be the enforcers of commanding right and forbidding wrong. One gay man who was kidnapped hoped that his kidnappers would not reveal his sexual orientation to his family, the shame of which would force him never to see them again. But hundreds of others have suffered a far worse fate at the hands of religious death squads that patrol the streets of Iraq's major cities looking for "effeminate men."

As reported by *Der Spiegel*, "In Baghdad a new series of murders began early this year, perpetrated against men suspected of being gay. Often they are raped, their genitals cut off, and their anuses sealed with glue. Their bodies are left at landfills or dumped in the streets." In the words of the head of Iraq's leading lesbian, gay, bisexual, and transgender organization, Iraq "is the most dangerous place in the world for sexual minorities." Even Turkey, where homosexuality is legal and where many Iraqis and Iranians ultimately flee, has seen a gay honor killing, which was carried out by one unfortunate young man's father. (There is of course a rampant hypocrisy at work here because there are significant gay and lesbian populations in all Islamic nations. Because affairs with women are so logistically difficult, for example, Arab men have long turned to other men to satisfy their sexual needs. In Afghanistan, too, wealthy tribesmen are known to purchase young boys for their personal pleasure.)

Many religions have difficulties with accepting homosexuality, needless to say. Some mainly Christian countries in

Africa have become appallingly homophobic in recent years. But even they do not prescribe the death sentence for gay people.

Honor Crimes in America

The practice of commanding right and forbidding wrong is not simply a problem for Muslim majority countries. It is increasingly a problem inside Muslim immigrant communities in the West.

I never cease to be amazed at how reluctant ordinary Americans are to believe that honor killings happen in the United States, too. In October 2009, for example, twenty-year-old Noor al-Maleki was killed by her father in suburban Phoenix, Arizona. He ran over her with his Jeep in a parking lot, crushing her body beneath its wheels. She did not die instantly, but lay gasping for breath as blood flowed from her mouth. What had she done in her father's eyes to merit such a death? The answer is that she liked makeup, boys, and Western music, and hoped to be able to support herself. She also refused to submit to the marriage her father had arranged for her to an Iraqi man who was in need of a green card. Noor wanted to choose her own fate. Instead, her father chose it for her. Others in the local Iraqi community defended Noor's father's actions. A thirty-something mother praying at a local mosque told *Time* magazine, as her daughter translated, "I think what he did was right. It's his daughter, and our religion doesn't allow us to do what she did."[14] (An Arizona jury found him guilty of second-degree murder and sentenced him to thirty-four years in prison.)

Or consider the case of the Egyptian-born taxi driver in

Dallas, Texas, who shot his seventeen- and eighteen-year-old daughters, Sarah and Amina, a total of eleven times for dating non-Muslim boys. At a vigil commemorating the two girls, their brother took the microphone and said: "They pulled the trigger, not my dad."[15] Or Fauzia Mohammed, who was stabbed eleven times by her brother in upstate New York because she wore "immodest clothing" and was "a bad Muslim girl." Or Aiya Altameemi, whose Iraqi-born father held a knife to her throat and whose mother and younger sister tied her to a bed and beat her because she was seen talking to a boy near their home in Arizona. Several months before, Aiya's mother had burned her face with a hot spoon because she refused to be married off to a man twice her age. Fauzia and Aiya survived, but they are scarred for life.

Similar crimes are being committed in Canada, too. The multimillionaire Afghan immigrant Muhammad Shafia killed his first wife and three daughters by locking them in a car and pushing it into a canal (the women may already have been drowned elsewhere) because the girls were becoming "too Westernized." Aqsa Parvez was a sixteen-year-old Toronto girl who wanted to be a fashion designer. Her father and brother strangled her to death for not wearing the hijab.

There can be no excuse for such foul acts. There can be no acceptable cultural defense. It should never be any woman's or girl's destiny to die at the hands of her own family—very often, in the documented American cases, her own father's—for the sake of some antiquated notion of family honor. Nor can any community be permitted to hush up the crime in the name of faith or cultural tradition.

In the West, honor violence is all too often conflated with domestic violence. Indeed, that is often how law enforcers and local media report cases of honor violence, sometimes out of a kind of self-censoring impulse. Underreporting of such cases encourages people to believe that honor violence "doesn't happen here" or, if it does, is no different from a drunkard punching his wife in the eye or menacing his son with a firearm.

But unlike domestic violence or abuse, where women and children (and sometimes also men) are nearly always brutalized in private, honor violence does not have to happen behind closed doors. Instead, the perpetrators often have the open support of family and community. There is no stigma because of the belief that the perpetrator is in the right. There is no need to leave bruises only where they will not show. Indeed, there can be social vindication and even redemption in a mutilated body, in a trail of blood. To escape a grisly death, a potential victim of honor violence must leave not only her abuser, but often her entire family and cultural community.

Whenever the apologists for honor violence say, "It is our religion," there must be an uncompromising reply: "Murder—and above all infanticide—cannot be sanctioned by any religion, by any faith, by any God."

Consider the case of the Pakistani man in Brooklyn who beat his wife to death with a stick because she made him a meal out of lentils rather than the goat meat he had requested. Though he was seventy-five and she was sixty-six, he left her body "a bloody mess." His defense attorney opened with the proposition that it was a culturally appropriate act because "he believed he had the right to hit and discipline his wife." At sentencing, the same attorney argued that prison would be a "hardship" because the man would not have access to Pakistani food. The New York judge sentenced the murderer to eighteen years to life.[16] But in a sharia zone, would the incident have even been reported, let alone come to trial?

Commanding Wrong

In 2010, in the British city of Derby, Kabir Ahmed and four other Muslim men passed out a leaflet entitled "Death Penalty?" and stuffed it through local mail slots. Illustrating the leaflet was a picture of a mannequin, hanging by a noose, with the message that homosexuality is punishable by death in Islam: "The death sentence is the only way this immoral crime can be erased from corrupting society and act as a deterrent for any other ill person who is remotely inclined in this bent way." It continued: "The only dispute amongst the classical authorities was the method employed in carrying out the penal code," and then went on to propose burning, being flung from a high point such as a mountain or building, or being stoned to death as suitable methods of death. Two other leaflets, entitled "Turn or Burn" and "God Abhors You," were also given out.

At his 2012 trial for stirring up hatred on the grounds of sexual orientation, Ahmed argued that he was in fact only spreading the word of God as taught through Islam: "My intention was to do my duty as a Muslim, to inform people of God's word and to give the message on what God says about homosexuality." According to the BBC, Ahmed also told the court he felt it was his duty as a Muslim to inform and advise people if they were committing sins, and that he would be failing if he did not. "My duty is not just to better myself but to try and better the society I live in," he added. "We believe we can't just stand by and watch somebody commit a sin, we must try and advise them and urge them to stay away from sin."[17]

Ahmed was sentenced to fifteen months in prison. After his release, he left his wife and three small children and joined IS. On November 7, 2014, he drove a truck laden with explosives into an Iraqi police convoy north of Baghdad, killing himself, an Iraqi general, and seven policemen, and injuring fifteen others.[18] A few months before he had told a *Newsweek* reporter, "It is for the sake of . . . religion and . . . honor. We are not for this life, but for the afterlife."[19] This is the doctrine of commanding right and forbidding wrong in action.

Ahmed's case is very far from unique. Consider this 2011 broadcast from a Muslim radio station in Leeds, England, during Ramadan. Speaking in Urdu, Rubina Nasir told listeners to Asian Fever's *Sister Ruby Ramadan Special*: "What should be done if they [practice homosexuality]? If there are two such persons among you, that do this evil, the shameful act, what do you have to do? Torture them; punish them; beat them and give them mental torture. Allah states, 'If they do such a deed, punish them, both physically and mentally. Mental punishment means rebuke them, beat them, humiliate them, admonish and curse them, and beat them up. This command was sent in the beginning because capital punishment had not yet been sent down.' "[20]

The following day Nasir was back on the air, talking about what happens when a Muslim man or woman gets married to a Mushrak—one who associates God with another (Jesus), i.e., a Christian.

> Listeners! Marriage of a Muslim man or woman with a Mushrak is the straight path to hellfire. Have my sisters and brothers, who live with people of bad religions or alien religions, ever thought about what would become

> of the children they have had with them—and the coming generation? Where the filth of *shirk* [the sin of following another religion] is present, where the dirt of *shirk* is present, where the heart is impure, how can you remove apparent filth? How many arrangements will you make to remove the apparent filth? We are saying that Mushraks have no concept of cleanliness and uncleanliness.[21]

For these comments, the radio station was fined £4,000 (around $6,000), but there was no move to suspend its broadcasting license.

Confronted with such flagrant acts of intolerance—such abuses of the freedom of speech—a free society must surely do more. For intolerance is the one thing a free society cannot afford to tolerate.

Only when Muslims—particularly those in Western countries—are free to say what they want, to pray or to not pray, to remain Muslim or to convert, or to have no faith at all; only when Muslim women are free to wear what they want, to go out as they want, to choose the partners that they want—only then will we be on a path to discover what is truly right and truly wrong in the twenty-first century. Commanding right and forbidding wrong are fundamentally at odds with the core Western principle of individual freedom. They, too, need to be removed from the central Islamic creed.

CHAPTER 7

JIHAD

Why the Call for Holy War Is a Charter for Terror

We don't expect Islamic holy war in Ottawa, Canada's chilly capital city. But in October 2014, a young Muslim named Michael Zehaf-Bibeau shot an unarmed Canadian soldier who was guarding the tomb of the unknown soldier at Ottawa's National War Memorial and then was himself killed in a shoot-out inside the Canadian Parliament's Hall of Honor. In the immediate aftermath, a *Washington Post* reader sent the following to the newspaper's website: "ISIL, via an incredible internet marketing, recruitment and promotion campaign, is delivering a message that is resonating with westerners. Western governments and society will need to figure out how and why this message of death is more appealing than the life these folks have been given in their countries."

That is the question, in various forms, that gets asked after each new atrocity, whether it happens in Oklahoma City or Sydney, Australia. In the wake of the shooting, stabbing, and attempted beheading of the British soldier Lee Rigby in broad daylight on a London street by two Muslim converts, the same question was asked. One of the men, Michael Adebolajo, gave his answer in a handwritten note he gave to a stunned bystander. The note read:

> To my beloved children know that to fight Allah's enemies is an obligation. The proofs of which are so numerous that but a handful of any of them cuts out the bewitching tongues of the Munafiqeen [hypocrites].
>
> Do not spend your days in endless dispute with the cowardly and foolish if it means it will delay you meeting Allah's enemies on the battlefield.
>
> Sometimes the cowardly and foolish could be those dearest to you so be prepared to turn away from them.
>
> When you set out on this path do not look left or right.
>
> Seek Shaheedala oh my sons . . .[1]

"Shaheedala" means martyrdom for the sake of Allah. It is the ultimate obligation—and reward—of the Islamic imperative of jihad: holy war.

The injunction to wage jihad is as old as the Qur'an, but in Muhammad's time there were no automatic weapons, no rocket-propelled grenades, no improvised explosive devices, no suicide vests. It was not possible to leave homemade bombs in backpacks near the finishing line of a race.

The carnage that erupted on April 15, 2013, some fifty yards from the finish line of the Boston Marathon, was apparently perpetrated by two brothers, Tamerlan and Dzhokhar

Tsarnaev. Born in the former Soviet Union to a Chechen father who had sought asylum in the United States in 2002, each of the brothers had received the gifts of free education, free housing, and free medical care from various U.S. governmental agencies. The younger brother, Dzhokhar, had already been granted his American citizenship, administered to him on, of all dates, September 11. Tamerlan was merely waiting for his final citizenship paperwork to be processed. The brothers spent months preparing for their bombing to take place on Patriots' Day, which commemorates the heroes of the American Revolution. How to explain such staggering ingratitude toward their adopted homeland?

Dzhokhar Tsarnaev offered at least the beginnings of an explanation in a note written not long before he was apprehended: "I'm jealous of my brother who ha[s] [re]ceived the reward of jannutul Firdaus [the highest level of Paradise] (inshallah) before me. I do not mourn because his soul is very much alive. God has a plan for each person. Mine was to hide in this boat and shed some light on our actions. I ask Allah to make me a shahied (iA) [a martyr inshallah] to allow me to return to him and be among all the righteous people in the highest levels of heaven. He who Allah guides no one can misguide. A[llah Ak]bar!"[2] He also offered this explicit account of his and his brother's motivations:

> the ummah is beginning to rise/ [unintelligible] has awoken the mujahideen, know you are fighting men who look into the barrel of your gun and see heaven, now how can you compete with that[?][3]

Dzhokhar Tsarnaev is very far from the only young man in the West to have fallen under the spell of jihad. Consider the near-perfect all-American life of Faisal Shahzad, a Pakistani national who also became a naturalized U.S. citizen. He arrived on a student visa, married an American, graduated from college, worked his way up the corporate ladder to become a junior financial analyst for a cosmetics company in Connecticut, and received his citizenship at the age of thirty. A year later, in 2010, Shahzad tried to blow up as many of his fellow citizens as possible in a failed car bombing in New York's Times Square. Prior to his courtroom sentencing, the criminal trial judge asked Shahzad about the oath of allegiance to the United States that he had taken, in which, like all newly minted citizens, he did "absolutely and entirely renounce and abjure all allegiance and fidelity to any foreign prince, potentate, state or sovereignty, of whom or which I have heretofore been a subject or citizen." Shahzad replied: "I sweared [*sic*], but I didn't mean it"—the legal equivalent of swearing with one hand and crossing his fingers with the other, but with far more damaging consequences. He then expressed his regret about the failure of his plot and added that he would gladly have sacrificed a thousand lives in the service of Allah. He concluded by predicting the downfall of his new homeland, the United States.

When trying to explain the violent path of some Islamists, Western commentators sometimes blame harsh economic conditions, dysfunctional family circumstances, confused identity, the generic alienation of young males, a failure to integrate into the larger society, mental illness, and so on. Some on the Left insist that the real fault lies with the mistakes of American foreign policy.

None of this is convincing. Jihad in the twenty-first century is not a problem of poverty, insufficient education, or any other social precondition. (Michael Zehaf-Bibeau was earning more than $90,000 a year working for a drilling company in British Columbia, where he also reportedly proclaimed his support of the Taliban and joked about suicide bombing vests, with no repercussions.) We must move beyond such facile explanations. The imperative for jihad is embedded in Islam itself. It is a religious obligation.

But it also reflects the influence of the strategic minds behind global jihad, in particular Sayyid Qutb, the author of *Milestones*, who explicitly argued that Islam was not just a religion but a revolutionary political movement; Abdullah 'Azzam, Osama bin Laden's mentor, who propounded an individualist "lone wolf" theory of jihad; and the Pakistani army general S. K. Malik, who argued in *The Quranic Conception of War* that the only center of gravity in warfare was the soul of the enemy and that therefore terror was the supreme weapon.[4]

In Great Britain, the radical cleric Anjem Choudary has declared: "We believe there will be complete domination of the world by Islam." That domination can only come through the waging of jihad. Through his words, Choudary has helped to send hundreds of Europeans to the battlefields in Iraq and Syria, as well as to plant the seeds for jihadist attacks inside Britain. Choudary also supports the IS beheadings of Americans and Britons, telling a *Washington Post* reporter that the victims deserved to die. This message may seem foreign or outlandish to most Westerners, but we underestimate its appeal at our peril.

The Call to Jihad

As a sixteen- and seventeen-year-old girl in Kenya, I believed in jihad. With the enthusiasm of idealistic young Americans who want to join the Peace Corps, I was ready for holy war.

For me, jihad was something to aspire to beyond chores for my mother and grandmother and my dreaded math class. The ideal of holy war encouraged me to get out of the house and engage in charitable work for others. It gave me a focus for my inner struggle; now I could struggle to be a better Muslim. Every prayer, every veil, every fast, every acknowledgment of Allah signaled that I was a better person or at least on the path to becoming one. I had value, and if the hardships of life in the Old Racecourse Road section of Nairobi felt overwhelming, it was only temporary. I would be rewarded in the afterlife.

That's how jihad is generally first presented to most young Muslims—as a manifestation of the inner struggle to be a good Muslim. It's a spiritual struggle, a path toward the light. But then things change. Gradually, jihad ceases to be simply an inner struggle; it becomes an outward one, a holy war in the name of Islam by an army of glorious "brothers" ranged against the enemies of Allah and the infidel. Yet this martial jihad seems even more appealing.

The origins of jihad can be traced back to the foundational Islamic texts.[5] Key verses in the Qur'an, and many verses in the hadith, call for jihad, a type of religious warfare to spread the land ruled by Allah's laws. For example:

- 9:5 "But when the forbidden months are past, then fight and slay the Pagans wherever ye find them, and seize them, beleaguer them, and lie in wait for them in every stratagem (of war); but if they repent, and establish regular prayers and practice regular charity, then open the way for them: for Allah is Oft-forgiving, Most Merciful."

- 8:60 "Against them make ready your strength to the utmost of your power, including steeds of war, to strike terror into (the hearts of) the enemies, of Allah and your enemies, and others besides, whom ye may not know, but whom Allah doth know. Whatever ye shall spend in the cause of Allah, shall be repaid unto you, and ye shall not be treated unjustly."
- 8:39 "And fight them on until there is no more tumult or oppression, and there prevail justice and faith in Allah altogether and everywhere."
- 8:65 "O Prophet! rouse the Believers to the fight. If there are twenty amongst you, patient and persevering, they will vanquish two hundred."

Today, these words have lost none of their appeal. Beguilingly presented by modern theorists of jihad such as Qutb, 'Azzam, and Malik, they can readily inspire young men to try to replicate the achievements of Muhammad's warriors in battle.

Celebrity Jihad

When I was a teenager, only a few decades ago, there were only so many jihadists who could be recruited. It was a tedious process of finding the right recruits in the right mosques and madrassas. It required a form of charismatic retail politics, of selecting, nurturing, and pulling along. Today, it is far easier. All a jihadist needs is access to a smartphone, and recruits will follow him. Twitter, Tumblr, Instagram, even the pages of Facebook have become virtual recruiting grounds with a global reach. For young people who have very limited chances to achieve fame and notoriety in their current situation, jihad is like one giant selfie. Suddenly, they have Twitter followers and video viewers. Suddenly, more and more people are paying attention to them. They become social media celebrities.

An Egyptian student, Islam Yaken, is a good example. He studied engineering, received a law degree, and was fluent in French and Arabic. A fitness buff who once posted workout tips and photos of his bare torso on his Facebook page, he left Egypt to join IS. His photo uploads changed from gym scenes to images of him riding a horse and holding a sword. The news raced across Egyptian social media websites, only amplifying his newfound celebrity.[6]

Jihadists do not have to wait for martyrdom to bring them fame. Thanks to electronic media, they can be immortalized in an instant. Photos and 140-character postings from Syria and Iraq currently litter the Internet. They show smiling jihadists, relaxed, with their rifles or trophies of war. A young man named Yilmaz, a Dutch national from a Turkish family, posted a photo of himself holding a cute Syrian toddler. After a Florida man, Moner Mohammad Abusalha, carried out a suicide bombing in Syria, an image of him smiling and holding a cat popped up online. Another who has achieved instant infamy is the man nicknamed Jihadi John, whose face was disguised but whose English accent was clearly audible as he appeared in IS videos with the severed heads of two American journalists and a British aid worker. As Shiraz Maher of the International Center for the Study of Radicalization at King's College, London, explains, the message is: "Come out here and have the time of your life. It makes it look like jihadi summer camp."

Jihad, it seems, has become a kind of hip lifestyle for disaffected youth. Online videos use "jihad rap." There is a distinctive jihadist look, too. In photos and videos, they all look the same: men in the backs of trucks, waving their rifles aloft, bearded, dressed in black. Whether they are IS warriors driving toward Baghdad, Boko Haram members striking a Christian village in northern Nigeria, Taliban fighters attacking a school in Peshawar, the style is very much the same.

Yet we should not confuse style with substance. While modern technology allows jihadist groups to glamorize their activities, the content of their videos remains firmly rooted in Islamic tradition and the theory of global jihad. These are rebels with a cause. In their own minds, they are reliving the glorious past of holy war, reenacting Muhammad's early battles against the Quraysh, when he and his men were grossly outnumbered yet still were victorious, egged on by Allah's promise of rewards for those who died as martyrs.

I was about eight years old when I first heard the tales of the Prophet's army, at my Qur'an school in Saudi Arabia. (Our teachers showed us dramatic video re-creations of the battles.) Make no mistake: today's jihadist fighters have been raised with these same stories—and often the ineptitude of the jihadists' opponents seems to make history repeat itself. In Iraq, government soldiers fled their positions when IS attacked, despite being better armed than their attackers. In Nigeria, too, despite substantial Western assistance, the authorities failed miserably to free "our girls" from Boko Haram.

After the U.S. consulate attack in Benghazi, Libya, and the airport attack in Karachi, Pakistan, the jihadist websites gloated that Allah had weakened the enemy, allowing victory—exactly the same story I heard from Somalis back in 1994 after eighteen American military personnel were killed and mutilated in Mogadishu. Even the release of Sergeant Bowe Bergdahl in Afghanistan in exchange for five Taliban leaders can be presented as another victory for Allah's warriors over the infidel.

The jihadists, then, are not simply disaffected youths from deprived backgrounds who have surfed the wrong websites. They are men and women with a sense of sacred mission. The words of a ten-year-old Palestinian boy, speaking after his father's own death, perfectly capture what I mean:

> By Allah, oh my father, I love you more than my own soul, but that is trivial because of my religion, my cause and my Al-Aqsa [the mosque in Jerusalem]. Father, my eyes will shed no tears, but my finger will pull the trigger—this trigger that I still remember. I will never forget, beloved father, the times when you taught me the love of jihad. You taught me the love of arms, so that I would be a knight, Allah willing. I will follow in your steps and fight the enemies on the battlefield. Every drop of blood that dripped from your pure body is worth dozens of bullets directed towards the enemies' chests. Tomorrow I will grow up, tomorrow I will avenge, and the battlefields will know who is the son of the Martyr, the commander, Ashraf Mushtaha. Finally, father, we are not saying goodbye, rather, I'll see you as a Shahid [Martyr] in Paradise. [I am] your son, who longs to meet you, the young knight, Naim, son of Ashraf Mushtaha.[7]

"You taught me the love of jihad." That is the message being heard today across the globe. And thousands are heeding it.

Global Jihad

The scale of the jihadist problem is growing much faster than most people in the West want to face. At the University of Maryland at College Park, the National Consortium for the Study of Terrorism and Responses to Terrorism (START), part of the Global Terrorism Database, tracks terror attacks worldwide. What they are finding is that "worldwide terrorism is reaching new levels of destructiveness," according to Gary LaFree, a START director and professor of criminology and criminal justice at Maryland. Leading this dramatic rise is an "incredible growth" in jihadist attacks perpetrated by "al-Qaeda affiliates." In 2012, START identified the six most lethal jihad terror groups as the Taliban (more than 2,500 fatalities), Boko Haram (more than 1,200), Al-Qaeda in the Arabian Peninsula (more than 960), Tehrik-e Taliban Pakistan (more than 950), Al-Qaeda in Iraq (more than 930), and Al-Shabaab (more than 700).

The numbers for 2013 and 2014 will likely be even higher. Places such as Iraq and Syria are of course a long way away from the United States: it is five and a half thousand miles from New York to Damascus. Even Europeans tend to regard the Middle East as distant: from London to Damascus is, after all, nearly three thousand miles.

To many of us, Syria may just seem like this decade's Bosnia or Rwanda; we tend to assume, in a slightly cynical or fatalistic way, that the next decade will bring along a new list of distant conflict zones. On an intellectual level, we may accept that we should be concerned about jihadists abroad, but on an emotional level, most people in the West are still disengaged.

But the rise of Western jihadists is changing that. Almost no one in the United States, Canada, Australia, or Europe could escape the ghastly spectacle of a British-born jihadist beheading helpless American and British captives.

A report from the AIVD, the Dutch intelligence service, describes a pattern that can be seen not only in the Netherlands but right across Western Europe: young Muslims are quickly moving from being merely "fellow traveler sympathizers" with jihadists to being fully fledged "ruthless fighters." It is not just an apostate like me who must now live in fear; even moderate Muslims face threats. "Muslims in the Netherlands who openly oppose joining the Syrian conflict and challenge the highly intolerant and antidemocratic dogma of jihadism have found themselves increasingly subject to physical and virtual intimidation," according to the AIVD.[8] High-profile Muslims who oppose the jihadists "cannot even go out in public without protection," while former Muslim radicals, who have turned away from the violent ideology, are severely threatened.[9] And the call to jihad is transmitted through multiple channels. As the AIVD report puts it: "it is now available in multiple forms and many languages, with material ranging from the movement's classic written works to sound recordings of lectures and films from the front line."[10]

The jihadists have the upper hand in Europe—and they know it. In April 2014, a Dutch jihadist addressed the following tweet directly to the AIVD: "Greetings from Syria! Intensively monitored for years, sent back 4 times and now drinking Pepsi in Syria? Que pasa what went wrong?" The AIVD report grimly predicts attacks throughout Europe, on governments, on Jews, on moderate Muslims, both Sunnis and Shiites. The threat, it concludes, is greater than ever before.[11]

Why should the United States be any different, even if in relative terms the Muslim share of the population is smaller than in most Western European countries? A Pew survey from 2007 noted that American Muslims under the age of thirty were twice as likely as older Muslims to believe that suicide bombings in defense of Islam could be justified, and 7 percent of American Muslims between the ages of eighteen and twenty-nine said that they had a "favorable" view of Al-Qaeda.[12]

While the proportion may be small, the absolute number of Americans committed to political Islam and willing to contemplate violence to advance its goals is not trivial. Another Pew survey, from 2011, found that somewhere around 180,000 American Muslims regarded suicide bombings as being justified in some way.[13] Abu Bakr al-Baghdadi, the leader of IS, is said to have told his U.S. Army Reservist guards when he walked away from four years of detention in Camp Bucca in Iraq, "I'll see you in New York." I fear it is only a matter of time before IS does indeed manifest itself in Manhattan.

Islam has always been transnational. It was founded and established and spread across the world when the nation-state and national identity were at best inchoate and more often nonexistent. People belonged to tribes, city-states, empires, or religious orders. But whereas Christianity was configured from its inception to co-exist with states and empires alike (if they would tolerate Christianity), Islam from the outset aspired to be church, state, and empire. If you are a self-respecting Islamist, you are therefore bound to be a crosser of national borders. You may need to gain local power, but your ultimate goal is to have Islam rule the world. And today you can write and talk openly about that goal on Facebook, Twitter, or wherever else you like.

Islamic State's social media mastermind is believed to be Ahmad Abousamra, a dual American-Syrian citizen, who grew up in the comfortable Boston suburb of Stoughton, while his father worked as an endocrinologist at Massachusetts General Hospital. He attended the private Xaverian Brothers Catholic high school in Westwood, Massachusetts, before transferring to Stoughton High in his senior year, when he made the honor roll. He also made the dean's list at Northeastern University.

If this sounds like a privileged upbringing, that's because it was. Yet, according to the testimony of FBI agents, Abousamra "celebrated" the 9/11 attacks and, while in college in the early 2000s, expressed his support for murdering Americans because "they paid taxes to support the government and were *kufar* [nonbelievers]." Abousamra worshipped at the same Cambridge mosque as the Tsarnaev brothers and five other high-profile terrorists, among them Afia Siddiqui, an MIT scientist turned Al-Qaeda agent known as "Lady Al-Qaeda," who was sentenced to eighty-six years in prison for planning a chemical attack in New York.

An MIT scientist. A dean's list student at Northeastern. These jihadists are hardly uneducated, unskilled, or impoverished. Some have been the beneficiaries of the best Western education that money can buy. That they have nevertheless committed themselves to holy war against the West is deeply perplexing to those of us who cannot imagine anything being more attractive than the Western way of life. That is why we cast around desperately for explanations of their behavior—any explanations, other than the obvious one.

The Roots of Jihad

In the immediate aftermath of the Boston Marathon bombings in 2013, there was a rush to deny that the Tsarnaev brothers had been motivated by religious radicalism. President Obama went out of his way to avoid referring to Islam in his statements after the Boston bombing. When it became impossible to deny that the perpetrators had in fact been avidly reading the online tirades of Abdullah Azzam, a Palestinian teacher and mentor of Osama bin Laden, the Islamic Society of Boston issued a bland statement saying that "one suspect [had] disagreed with the moderate American-Islamic theology of the ISB Cambridge mosque."

It was much the same story just over a month later, on May 22, when Lee Rigby was hacked to death in Woolwich. Within hours, a woman named Julie Siddiqi, representing the Islamic Society of Britain (and a convert to the faith), stepped before the microphones to attest that all good Muslims were "sickened" by the attack, "just like everyone else." *The Guardian* ran a headline quoting a Muslim Londoner: "These poor idiots have nothing to do with Islam." Try telling that to Lee Rigby's murderer who killed him while yelling *"Allahu akbar"* (God Is Great).

Omar Bakri also claimed to speak for the true faith following the Woolwich killing. Of course, he was unavailable for the cameras in England because the Islamist group he founded, Al-Muhajiroun, was banned in 2010, so he spoke from Tripoli in northern Lebanon, where he now lives under an agreement with the Lebanese government that prevents him from leaving the country for thirty years. A decade earlier, in London, Bakri had taught Michael Adebolajo, the accused Woolwich killer who was videotaped at the scene. "A quiet man, very shy, asking lots of questions about Islam," Bakri recalled of his student, the terrorist. The teacher was impressed to see in the grisly video of Lee Rigby's murder how far his shy disciple had come, "standing firm, courageous, brave. Not running away. . . . The Prophet said an infidel and his killer will not meet in Hell. That's a beautiful saying. May God reward him for his actions. . . . I don't see it as a crime as far as Islam is concerned."[14]

Omar Bakri is not making up Muhammad's words. If the Qur'an or the hadith urges the believer to kill infidels ("slay them wherever ye catch them" [2:191]) or to behead them ("when ye meet the Unbelievers [in fight], smite at their necks; At length, when ye have thoroughly subdued them, bind a bond firmly [on them]" [47:4])—or to whip adulterers and stone them to death (Sahih Muslim 17:4192), then we cannot be wholly surprised when fundamentalists do precisely those things. Those who say that the butchers of Islamic State are misinterpreting these verses have a problem. The Qur'an itself explicitly urges pitilessness.

Or consider the case of Boko Haram, the organization that briefly attracted the attention of the American public by kidnapping 276 schoolgirls in Nigeria last year. The translation of Boko Haram from the Hausa language is usually given in English-language media as "Western Education Is Forbidden." But "Non-Muslim Teaching Is Forbidden" might be more accurate. Like individual terrorists, organizations such as Boko Haram do not spring from nowhere. The men who establish such groups, whether in Africa, Asia, or even Europe, are members of long-established Muslim communities, most of whose members are happy to lead peaceful lives. To understand why the jihadists are flourishing, you need to understand the dynamics within those communities.

It begins simply enough, usually with the establishment of an association of men dedicated to the practice of the *sunnah* (the tradition of guidance from the Prophet Muhammad). There will be a lead preacher, not unlike Boqol Sawm, the Muslim Brotherhood imam I encountered as a girl in Nairobi. Much of the young man's preaching will address the place of women. He will recommend that girls and women be kept indoors and covered from head to toe if they are to venture outside. He will also condemn the permissiveness of Western society.

What kind of response will he encounter? In the United States and in Europe, some moderate Muslims may quietly draw him to the attention of authorities. Women may voice concerns about the attacks on their freedoms. But in other parts of the world, where law and order are lacking, such young men and their extremist messages can thrive. In particular, where governments are weak, corrupt, or nonexistent, the message of Boko Haram and its counterparts is especially compelling. Not implausibly, they can blame poverty on official corruption and offer as an antidote the pure principles of the Prophet.

But why do so many young men turn from these words to violence? At first, they can count on some admiration for this fundamentalist message from within their own communities. Some may encounter opposition from established Muslim leaders who feel threatened. But the preacher and his cohorts persevere because perseverance in the *sunnah* is one of the most important keys to heaven. And over time, the following grows, to the point where it is as large as that of the Muslim community's established leaders. That is when the showdown happens—and the argument for "holy war" suddenly makes sense to leader and follower alike.

The history of Boko Haram has followed precisely this script. The group was founded in 2002 by a young Islamist called Mohammed Yusuf, who started out preaching in a Muslim community in Borno state of northern Nigeria. He set up an educational complex, including a mosque and an Islamic school. For seven years, mostly poor families flocked to hear his message. But in 2009, the Nigerian government investigated Boko Haram and ultimately arrested several members, including Yusuf himself. The crackdown sparked violence that left about seven hundred dead.

Yusuf soon died in prison—the government said he was killed while trying to escape—but the seeds had been planted. Under one of Yusuf's lieutenants, Abubakar Shekau, Boko Haram turned to jihad. In 2011, Boko Haram launched its first terror attack in Borno. Four people were killed, and from then on violence became an integral part, if not the central part, of its mission.

It is no longer plausible to argue that organizations such as Boko Haram—or, for that matter, Islamic State—have nothing to do with Islam. It is no longer credible to define "extremism" as some disembodied threat, meting out death without any ideological foundation, a problem to be dealt with by purely military methods, preferably drone strikes. We need to tackle the root problem of the violence that is plaguing our world today, and that must be the doctrine of Islam itself.

The Practice of Jihad: The Worldwide War on Christians

One of the most devastating manifestations of the modern era of jihad is the violent oppression of Christian minorities in Muslim-majority nations all over the world.

In Islamic history, the land controlled by Islam is referred to as *dar al-Islam* (the abode of Islam). The land controlled by non-Muslims is *dar al-harb* (the abode of war).[15] Historically, after being conquered by Muslims, groups deemed People of the Book, including Jews, Christians, and Zoroastrians, were required to pay a special tax, the *jizya*, as a mark of their humiliation. If they did so, they were allowed to keep their religion (9:29). Yet there was always a strain of "eliminationism" in Islam, too. The Prophet himself promised to "expel the Jews and Christians from the Arabian Peninsula and . . . not leave any but Muslims" (Sahih Muslim 19: 4363–67). The Qur'an (5:51) warns Muslims: "take not the Jews and the Christians for your friends and protectors." Muslim men may marry Jewish or Christian women but Muslim women may not marry non-Muslim men because under Islamic law the religious identity of children is passed through the father (5:5).

Modern Islamists go further. In some countries, governments and their agents openly sponsor anti-Christian violence, burning churches and imprisoning observant Christians. In others, rebel groups and self-proclaimed vigilantes have taken matters into their own hands, murdering Christians and driving them from regions where their roots go back centuries. Often, local leaders and governments do little to stop them or simply turn a blind eye.

This phenomenon of Christophobia (as opposed to the far more widely discussed "Islamophobia") receives remarkably little coverage in the Western media. Part of this reticence may be due to fear of provoking additional violence. But part is clearly a result of the very effective efforts by lobbying groups such as the Organization of Islamic Cooperation and the Council on American-Islamic Relations. Over the past decade, these and similar groups have been remarkably successful in persuading journalists and editors in the West to think of each and every example of perceived anti-Muslim discrimination as an expression of a deep-rooted Islamophobia. This, of course, extends with an Orwellian illogic to coverage of Muslim violence against Christians. Yet any fair-minded assessment of recent events leads to the conclusion that the scale and severity of Islamophobia pales in comparison with the Christophobia evident in Muslim-majority nations from one end of the globe to the other.

Take Nigeria, where the population is almost evenly split between Christians and Muslims, who for years have lived on the edge of civil war. But the stakes have risen dramatically with the gains made by Boko Haram, which has openly stated that it will kill all of Nigeria's Christians. And it is making good on its promise. In the first half of 2014, Boko Haram killed at least 2,053 civilians in ninety-five attacks.[16] They have used machetes, guns, and gasoline bombs, shouting "*Allahu akbar*" (God is great) while launching their attacks, one of which—on a Christmas Day gathering—killed forty-two Catholics. They have targeted bars, beauty salons, and banks. They have murdered Christian clergymen, politicians, students, policemen, and soldiers.

In Sudan, the authoritarian government of the Sunni Muslim north of the country has for decades tormented Christian (as well as animist) minorities in the south. What has often been described as a civil war is in practice the Sudanese government's sustained policy of persecution, which culminated in the infamous genocide in Darfur that began in 2003. Even though Sudan's Muslim president, Omar al-Bashir, has been charged at the International Criminal Court in The Hague with three counts of genocide, and despite the euphoria that greeted South Sudan's independence in 2012, the violence has not ended. In South Kordofan, for example, Christians are still subjected to aerial bombardment, targeted killings, the kidnapping of children, and other atrocities. Reports from the United Nations indicate that there are now 1 million internally displaced persons in South Sudan.[17]

Both kinds of persecution—undertaken by nongovernmental groups as well as by agents of the state—have come together in Egypt in the aftermath of the Arab Spring. On October 9, 2012, in the Maspero area of Cairo, Coptic Christians—who make up roughly 5 percent of Egypt's population of 81 million[18]—marched in protest against a wave of attacks by Islamists, including church burnings, rapes, mutilations, and murders, that followed the overthrow of Hosni Mubarak's dictatorship. During the protest, Egyptian security forces drove their trucks into the crowd and fired on protesters, crushing and killing at least twenty-four and wounding more than three hundred people.[19] Within two months, tens of thousands of Copts had fled their homes in anticipation of more attacks.[20]

Nor is Egypt the only Arab country where Christian minorities have come under attack. Even before the advent of IS, it was dangerous to be a Christian in Iraq. Since 2003, more than nine hundred Iraqi Christians (most of them Assyrians) have been killed in Baghdad alone, and seventy churches have been burned, according to the Assyrian International News Agency (AINA). Thousands of Iraqi Christians have fled as a result of violence directed specifically at them, reducing the number of Christians in the country from just over a million before 2003 to fewer than half a million today. AINA understandably describes this as an "incipient genocide or ethnic cleansing of Assyrians in Iraq." The recent decimation by IS forces of Mosul's two-thousand-year-old Christian population—who fled under threat of death or forced conversion, and saw their possessions stolen and looted, their homes marked with "N" (for Nazarene) and their churches desecrated—is merely the latest episode in a campaign of persecution.

One Mosul resident, Bashar Nasih Behnam, escaped with his two children. "There is not a single Christian family left in Mosul," he said. "The last one was a disabled Christian woman. They came to her and said you have to get out and if you don't we will cut off your head with a sword. That was the last family." Those fleeing were also robbed: the IS fighters took their money and gold, ripped earrings from women's ears, and confiscated mobile phones.

Then there are the states where intolerance is part and parcel of the nation's legal code. Pakistan's Christians are a tiny minority—only about 1.6 percent of a population of more than 180 million. But they are subject to intense segregation and discrimination: allowed to shop only at a few sparsely stocked stores, forbidden to draw water from wells earmarked for Muslims, and forced to bury their dead, stacked on top of one another, in tiny graveyards because Muslims cannot be buried near people of other faiths.

They are also subjected to Pakistan's draconian blasphemy laws, which make it illegal to declare belief in the Christian Trinity. When a Christian group is suspected of transgressing the blasphemy laws, the consequences can be brutal. In the spring of 2010, the offices of the international Christian aid group World Vision were attacked by ten men armed with grenades, who left six people dead and four wounded. A militant Muslim group claimed responsibility for the attack, on the ground that World Vision was working to subvert Islam. (In fact, it was helping the survivors of a major earthquake.)

Not even Indonesia—often touted as the world's most tolerant, democratic, and modern majority-Muslim nation—has been immune to the fever of Christophobia. Between 2010 and 2011, according to data compiled by the *Christian Post*, the number of violent incidents committed against religious minorities (and at 8 percent of the population, Christians are the country's largest minority) increased by nearly 40 percent, from 198 to 276.

Despite the fact that more than a million Christians live in Saudi Arabia as foreign workers, even private acts of Christian prayer are banned. To enforce these totalitarian restrictions, the religious police regularly raid the homes of Christians and bring them up on charges of blasphemy in courts where their testimony carries less legal weight than a Muslim's. Saudi Arabia bans the building of churches, and its textbooks enshrine anti-Christian and anti-Jewish dogma: sixth-grade students are taught that "Jews and Christians are enemies of the believers." An eighth-grade textbook says, "The Apes are the people of the Sabbath, the Jews; and the Swine are the infidels of the communion of Jesus, the Christians."[21] Even in Ethiopia, where Christians make up a majority of the population, church burnings by members of the Muslim minority have become a problem.

Anti-Christian violence is not centrally planned or coordinated by some international Islamist agency. It is, rather, an expression of anti-Christian animus that transcends cultures, regions, and ethnicities. As Nina Shea, director of the Hudson Institute's Center for Religious Freedom, pointed out in an interview with *Newsweek*, Christian minorities in many majority-Muslim nations have "lost the protection of their societies."

Of course, intolerance of different faiths is not unique to Islam. The Roman Empire first persecuted Christians, then persecuted non-Christians after Christianity was adopted as the Empire's official religion. In medieval Christendom there was no "religious freedom" as we would recognize it today; heretics were cruelly punished, Jews persecuted. When Pope Urban II called for the first crusade in 1095, he told knights willing to journey to Jerusalem that they would be forgiven all their past sins if they killed unbelievers in the Holy Land. And when European Christians set out to conquer and colonize the world, their treatment of "heathens" was often brutal to the point of genocide. Yet Patricia Crone argues that there was always something unique about the Muslim concept of jihad—"the belief that God had chosen one people over others and ordered them to go conquer the earth." Christians today, with few exceptions, repudiate the intolerance of the past. In the twentieth century, the horrors of the Holocaust forced Christian thinkers to confront the pernicious role of anti-Semitism in European history. The contrast with the Muslim world is stark. There, intolerance is on the rise and the remit of jihad has been extended to include all nonbelievers.

Why Are the Jihadists Winning? Because We Are Letting Them

In July 2014, the prospect of a flag bearing the words of the Shahada being raised over Downing Street got the attention of one hundred British imams, who signed a letter urging "British Muslim communities not to fall prey to any form of sectarian divisions or social discord" but rather "to continue the generous and tireless efforts to support all of those affected by the crisis in Syria and unfolding events in Iraq . . . from the UK in a safe and responsible way." Qari Muhammad Asim, the imam at the Makkah Mosque in Leeds and one of the authors of the letter, told BBC radio: "Imams from a cross-section of theological backgrounds have come together to give a very strong message to young British Muslims who might be inclined to go to Syria or Iraq to fight, saying to them, 'Please don't expose yourselves, don't put your lives at risk and the lives of others around you.'" Responding to a question, he went further:

> Islam itself has been hijacked and [some] people . . . have been completely brainwashed. It's completely ridiculous to say that people, fellow human beings, are enemies and as a result they should be blown up. Obviously, social media plays a huge part, the Internet plays a huge part, in brainwashing and radicalizing people.[22]

According to Asim, more than one hundred imams were planning to launch appeals on social media and platforms like Twitter. They have even developed a website, imamsonline.com.

"A lot of work needs to be done," he acknowledged. But "it's not just the responsibility of the Muslim community and the imams. It's law enforcement, intelligence services. We all need to work together in partnership and make sure that young British Muslims are not preyed upon by those who want to use them for their own political gains."

It would, of course, be deeply reassuring if we could believe that the Western jihadists are merely the victims of online brainwashing and that a few moderate websites would soon fix the problem. But the reality is very different. Those who have been recruited to the cause of jihad have not just been unlucky in their Internet browsing selections. Since the 1990s, foreign-born imams have established themselves in pockets of London and other major European cities, preaching sermons and distributing audio recordings in which they have explicitly and repeatedly called for jihad.

With the best of intentions, no doubt, the British government opened its doors to many of these imams, often recognizing them as legitimate asylum seekers and offering them the usual welfare benefits available to those fleeing persecution. To give just one example, the Finsbury Park Mosque, led by the Egyptian imam and now convicted terrorist Abu Hamza al-Masri, had among its congregation the "shoe bomber" Richard Reid, the 9/11 "twentieth hijacker" Zacarias Moussaoui, the would-be Los Angeles airport bomber Ahmed Ressam, as well as Ahmed Omar Saeed Sheikh, who stands accused by the Pakistani government of murdering the *Wall Street Journal* reporter Daniel Pearl.

In response to this kind of threat, the British government developed what it calls the "Prevent strategy." Prevent is supposed to stop Britons and residents from being drawn into terrorist activities and networks, by working with all branches of government, from education to law enforcement. For instance, Prevent is supposed to help the immigration authorities to deny visas to extremist imams. But the remit of Prevent is broad: it is supposed to cover all forms of terrorism, from right-wing extremism to something vaguely called "nonviolent extremism," whatever that means.

The potential weaknesses of this approach can be seen in the comments of one of its regional managers, Farooq Siddiqui, who in 2014 used a Facebook chat to offer his approval to Britons who wanted to travel to Syria to fight against the regime of President Assad, saying that these men had "walked the walk." He compared these fighting jihadists to British Jews who might join the Israel Defense Forces and could then return to the United Kingdom, arguing on that basis that jihadists returning from Syria should not face automatic arrest. "If a man describes himself as wanting to help the oppressed and dies," Siddiqui wrote, "in that case he is a martyr."[23] It is not immediately obvious what a man like Siddiqui is going to prevent, aside from a serious discussion of the problem Britain faces.

Ghaffar Hussein, the managing director of Quilliam, a British think tank working on combatting terrorism, notes that jihad is appealing because of its "one size fits all" set of answers to complex problems. Introspection is not required, he notes, because all blame is shifted to outside enemies and "anti-Muslim conspiracy theories." The jihad narrative has therefore become "the default anti-establishment politics of today. It is a means of expressing solidarity and asserting a bold new identity while being a vehicle for seeking the restoration of pride and self-dignity." In response, "mainstream Muslim commentators"—not to mention non-Muslims—have failed to articulate a positive narrative that does not simply reinforce the idea that Muslims are somehow victims. In short, Hussein's argument is that the jihadists have the more compelling narrative. To understand the power of that narrative, let's look more closely at what motivates young Western-educated Muslims to sign up for jihad.

In 2013 Umm Haritha, a twenty-year-old Canadian, traveled to Syria via Turkey to join Islamic State. Within a week, she had married an IS fighter, a Palestinian national who had been living in Sweden. He was killed five months later and Umm, a widow, turned to blogging, offering advice to others who wished to move to Syria, marry jihadists, and create families inside the IS caliphate.

Her words make for interesting reading. In an interview with Canada's CBC via text messages, Umm described herself as "middle class," adding that her decision to join jihad was made by a desire to "live a life of honor" under Islamic law rather than the laws of the "*kufar*," or unbelievers. She had begun her journey to jihad in Canada, where she donned the niqab, a veil that exposes nothing more than the wearer's eyes. She told her interviewer that she felt "mocked" and harassed by her fellow Canadians, adding, "Life was degrading and an embarrassment and nothing like the multicultural freedom of expression and religion they make it out to be, and when I heard that the Islamic State had sharia in some cities in Syria, it became an automatic obligation upon me since I was able to come here."[24]

Umm's online postings describe life in Manbij, an IS-controlled city of 200,000 close to the Turkish border, and show images such as the white loudspeaker van that patrols the city streets to remind residents of their daily prayers. She notes approvingly that a man was recently crucified and beheaded for the crime of robbing and raping a woman. And she adds that many of those who have moved to the caliphate have "ripped up their passports." Abu Bakr al-Baghdadi, the IS leader, who has renamed himself "Caliph Ibrahim," has called on Muslims worldwide to move to the caliphate, saying, "Those who can immigrate to the Islamic State should immigrate, as immigration to the house of Islam is a duty." As the stepbrother of a radicalized British man explained, the purveyors of jihad know what their recruits "are craving—identity, respect, empowerment. They push all the right buttons—make them feel special. And once you're in the door, it's like family. They look after each other."

Consider, too, a 2014 BBC 5 Live interview with a man calling himself Abu Osama, who claimed to be from the north of England and said that he was training with the Al-Nusra Front in Syria with the ultimate goal of establishing a caliphate (*Khilafah* in Arabic) across the Islamic world. Osama told the BBC: "I have no intention of coming back to Britain, because I have come to revive the Islamic Khilafah. I don't want to come back to what I have left behind. There is nothing in Britain—it is just pure evil." And for emphasis he added: "If and when I come back to Britain it will be when this *Khilafah*—this Islamic state—comes to conquer Britain and I come to raise the black flag of Islam over Downing Street, over Buckingham Palace, over Tower Bridge and over Big Ben."[25] (Anjem Choudary has promised the same, predicting that the black flag of IS will fly over both 10 Downing Street and the White House after the conclusion of the great global battle that is now under way.)

Such seemingly wild narratives are not out of the mainstream; rather, they present jihad just in the way it has always been taught. "If you look at the history of Islam," as the young jihadist Osama put it, "you will see that the Prophet fought against those who fought against him. He never fought those that never fought against the Islamic state. Where I am, the people love us, the people love the mujahideen, the warriors." As for Osama's family, at first they had found it "hard to accept," but he had won them over to his "good cause." As he put it: "They are a bit scared but I tell them we will meet in the afterlife. This is just a temporary separation. They said, 'We understand now what you are doing,' and my mother said, 'I have sold you to Allah. I don't want to see you again in this world.'"[26]

Is Jihadism Curable?

The Harvard Kennedy School scholar Jessica Stern has spent years studying counterterrorism and, in particular, efforts to prevent the spread of jihad. Indeed, she was consulted on the development of an anti-jihad effort in the Netherlands after the brutal murder of Theo van Gogh ten years ago. In a recent article, she describes in detail a Saudi Arabian jihadist rehabilitation program that has "treated" thousands of militants, and claims that the graduates have been "reintegrated into mainstream society much more successfully than ordinary criminals."[27]

The Saudi approach, Stern notes, is inspired by the efforts of other governments in other regions of the world to "deprogram" everyone from neo-Nazis to drug lords. The goal is to get them "to abandon their radical ideology or renounce their violent means or both." The method is a full-time residential program that includes "psychological counseling, vocational training, art therapy, sports, and religious reeducation," along with "career placement" services for themselves and their families, if needed. Upon completion, the program's graduates—some of whom have been previously incarcerated in the U.S. detention center at Guantánamo Bay—receive housing, a car, and even funds to pay for a wedding. The Saudis will even assist them with finding a wife.

But the program doesn't end there. There is what Stern describes as "an extensive post-release program as well, which involve[s] extensive surveillance." Rather like convicted sex offenders in the West, ex-jihadists will be monitored for most if not all of the rest of their lives. Stern goes on to explain that the "guiding philosophy" behind the program is that "jihadists are victims, not villains, and they need tailored assistance." Accordingly, the Saudis have a very specific term for the program's participants. They are "beneficiaries."

Stern maintains that, while terrorist movements "often arise in reaction to an injustice, real or imagined," that the supporters "feel must be corrected," ideology generally plays a limited role in someone's decision to join the terror cause. She writes: "The reasons that people become terrorists are as varied as the reasons that others choose their professions: market conditions, social networks, education, individual preferences. Just as the passion for justice and law that drives a lawyer at first may not be what keeps him working at a law firm, a terrorist's motivations for remaining in, or leaving, his 'job' change over time." Stern also argues that the terrorists who "claim to be driven by religious ideology are often very ignorant of Islam." The Saudi "beneficiaries" have, she writes, little in the way of formal education and a limited understanding of Islam.

I am deeply skeptical about all this, for two reasons. First, as part of the Saudi program Stern describes, clerics are brought in to teach the beneficiaries that only "the legitimate rulers of Islamic states, not individuals such as Osama bin Laden, can declare a holy war. They preach against *takfir* [accusing other Muslims of apostasy] and the selective reading of religious texts to justify violence." One participant in the program told her: "Now I understand that I cannot make decisions by reading a single verse. I have to read the whole chapter." No matter how well intentioned this approach may be, it leaves the core concept of jihad intact.

Second, we should not forget that the global jihadist network would not exist on anything like the scale it does today if it had not been for Saudi funding—to say nothing of the millions that have flowed to terrorist organizations from other Gulf states. As Nabeel al-Fadhel, a liberal member of Kuwait's Parliament, told *The Christian Science Monitor*, "There isn't a bomb that explodes anywhere [inside Syria] without some of its material financed by Kuwait." Noting the vast number of Kuwaitis who have donated to the jihadist cause, he added that while they may "think they are getting closer to God by giving this money," instead, "it is going to places [they] never dreamt of."[28]

The last people we should expect to develop an effective counterforce to jihad are the rulers of those countries that, over the past thirty years, have played the biggest role in funding the Medina Muslims who have been jihad's most ardent advocates.

Decommissioning Jihad

In one of the many IS videos that can be found online, a British man who identifies himself as Brother Abu Muthanna al Yemeni extolls the virtues of jihad. He encourages foreign Muslims "to answer the call of Allah and His Messenger when He calls you to what gives you life. . . . What He says gives you life is jihad."[29] This is not empty rhetoric. We need to answer these words. We need more than just a counternarrative. We need a theological reply.

The nuclear arms race of the Cold War was not won by the proponents of unilateral disarmament. No matter how many thousands of people turned out for antinuclear marches in London or Bonn, missiles were still deployed in NATO countries and pointed at the Warsaw Pact countries, which had their own missiles pointed right back at the West. The only way the arms race ended was with the ideological and political collapse of Soviet communism, after which there was a large-scale (though not complete) decommissioning of nuclear weapons. In much the same way, we need to recognize that this is an ideological conflict that will not be won until the concept of jihad has itself been decommissioned. We also have to acknowledge that, far from being un-Islamic, the central tenets of the jihadists are supported by centuries-old Islamic doctrine.

The IS spokesman Abu Muhammad al-Adnani recently called on Muslims to use all means to kill a "disbelieving American or European—especially the spiteful and filthy French—or an Australian or a Canadian."[30] "Please don't" is not an adequate reply. As Ghaffar Hussain, himself a former

Islamist, has said, "You need to stand up, challenge them, and rubbish their ideas."

It is obviously next to impossible to redefine the word "jihad" as if its call to arms is purely metaphorical (in the style of the hymn "Onward Christian Soldiers").[31] There is too much conflicting scripture, and too many examples from the Qur'an and hadith that the jihadists can cite to bolster their case.

Therefore I believe the best option would be to take it off the table. If clerics and imams and scholars and national leaders around the world declared jihad "*haram*," forbidden, then there would be a clear dividing line. Imagine the impact if those hundred imams in Great Britain had explicitly renounced the entire concept of jihad. Imagine if the kingdom of Saudi Arabia, home to Islam's holy shrines, itself renounced jihad, rather than turning the jihadists into beneficiaries of (yet more of) its largesse.

And if that is too much to expect—if Muslims simply refuse to renounce jihad completely—then the next best thing would be to call their bluff about Islam being a religion of peace. If a tradition truly exists within Islam that interprets jihad as a purely spiritual activity, as Sufi Muslims tend to do, let us challenge other Muslims to embrace it. Christianity was itself once a crusading faith, as we have seen, but over time it abandoned its militancy. If Islam really is a religion of peace, then what is preventing Muslims from doing the same?

CHAPTER 8

THE TWILIGHT OF TOLERANCE

The first time I stood up to speak in a public setting was shortly after September 11, 2001. It was a public forum, a "discussion house," which is a relatively common institution in the Netherlands. I was working at a small but well respected social democratic think tank, and my boss suggested that I go.

The discussion was being hosted by a Dutch newspaper, a publication that was originally religious (Protestant), but now was very secular, and the topic was "Who Needs a Voltaire, the West or Islam?" The auditorium was packed to capacity. People who couldn't find seats were standing along the walls. And in many ways it was an interesting and unusual gathering because there were so many Muslim participants in the audience. Normally these things were almost all white because the discussion topics would be things like "How Much Control Do We Cede to the European Union?" or "Why Should

We Give Up the Guilder for the Euro?" On this night, however, the usual members of the Amsterdam elite were rubbing shoulders with Muslims from Turkey, Morocco, and other nations, nearly all of them immigrants or the children of immigrants to the Netherlands.

There were six speakers for the evening, and five of them essentially said that it was the West that needed a Voltaire, meaning that the West was the place most in need of reform. Their argument was that the West had a blind spot, that it had a long and wicked history of exploitation and imperialism, that it was tone deaf to what went on in the rest of the world, and it needed another Voltaire to explain all of this.

I was sitting in the middle of this sea of faces, white, brown, and black, and just listening, increasingly aware that I disagreed with what was being said. Finally, the sixth panelist spoke, a man from Iran, a refugee, a lawyer. "Well," he said, "look at these people in this room. The West has not one Voltaire, but thousands if not millions of Voltaires. The West is used to criticism, it's used to self-criticism. All the sins of the West are out there for everyone to see." And then he said: "It's Islam that needs a Voltaire." He discussed a list of all the things that are wrong or questionable about Islam—points that resonated with me. And for this he was booed; he was shouted down. (Ironically, ten years later, Irshad Manji, a staunch advocate of Islamic reform, spoke in this same hall. By then, the crowd had completely changed. It was packed not with curious observers, but with hard-line, fundamentalist Islamists, and that night the audience grew so combative that Irshad had to be hustled out by security.)

After the Iranian lawyer spoke, there was a break, and then the audience was given a chance to ask questions. I waved my hand, and someone with the microphone saw my black face and probably thought, "for the sake of diversity"—the white organizers of such events were in fact quite keen to hear what went on in the heads, households, and communities of immigrants. He gave me the microphone. I stood up and agreed with the Iranian. I said: "Look at you guys. There are six people there, you've invited six speakers, and one of them is the Voltaire of Islam. You guys have five Voltaires, just allow us Muslims one, please." That led a newspaper editor to ask me to write an essay, to which he gave the headline "Please Allow Us One Voltaire."

In the months and years that followed, I read more and more widely. I read Western views of Islam and Muslim culture. I read more Western liberal thinkers. And I read about the Muslim reformers of the past. My conclusion remains that Islam still needs a Voltaire. But I have come to believe it is in dire need of a John Locke as well. It was, after all, Locke who gave us the notion of a "natural right" to the fundamentals of "life, liberty, and property." But less well known is Locke's powerful case for religious toleration. And religious toleration, however long it took to be established in practice, is one of the greatest achievements of the Western world.

Locke made the case that religious beliefs are, in the words of the scholar Adam Wolfson, "matters of opinion, opinions to which we are all equally entitled, rather than quanta of truth or knowledge."[1] In Locke's formulation, protection against persecution is one of the highest responsibilities of any government or ruler. Locke also argued that where there is coercion and persecution to change hearts and minds, it will "work" only at a very high human cost, producing in its wake both cruelty and hypocrisy. For Locke, no one person should "desire to impose" his or her view of salvation on others. Instead, in his vision of a tolerant society, each individual should be free to follow his or her own path in religion, and respect the right of others to follow their own paths: "Nobody, not even commonwealths," Locke wrote, "have any just title to invade the civil rights and worldly goods of each other upon pretense of religion."[2]

What is often forgotten is that Locke restricted this freedom of religion to various Protestant denominations. He did not include the Roman Catholic Church because "all those who enter into it do thereby *ipso facto* deliver themselves up to the protection and service of another prince." Were Locke alive today, I suspect he would make a similar argument about Islam. So long as there are some Muslims who regard Muhammad's teachings in Medina as trumping their loyalty to the states of which they are citizens, there will be a legitimate suspicion that tolerance of Islam endangers the security of those states. The central question for Western civilization remains what it was in Locke's day: What exactly can we *not* tolerate?

Let us begin with the oppression of half of humanity.

Rights in Retreat

Today, more than two hundred years after Voltaire and three hundred years after John Locke, the rights of women are in retreat throughout the Muslim world. Consider, by way of a simple illustration, the way that Muslim women are permitted to dress. It is not the most important human right, I admit. But it is a freedom most women care about.

Look at photographs of any of the Muslim cities of the world in the 1970s: Baghdad. Cairo. Damascus. Kabul. Mogadishu. Tehran. You will see that very few women in those days were covered. Instead, on the streets, in office buildings, in markets, movie theaters, restaurants, and homes, most women dressed very much like their counterparts in Europe and America. They wore skirts above the knee. They wore Western fashions. Their hair was done up and visible.

Today, by contrast, a mere photo of a woman walking on the streets of Kabul with a knee-length skirt becomes a viral happening on the Internet, and sparks widespread condemnation as "shameful" and "half-naked," with the government criticized for "sleeping." When I was a girl in primary school in Nairobi, those who covered their heads were the exceptions—fewer than half of all the girls. A few years ago, I googled my old primary school. In the photos posted, nearly every girl was covered.

This is not just about how we dress. If you are a woman living in Saudi Arabia, you want to drive, you want to go out of the house without a male guardian. You may well have money, but you have nothing to do except sit at home or shop under male supervision. In Egypt, you are battling against a rising tide of sexual harassment—99 percent of women report being sexually harassed and up to eighty sexual assaults occur in a single day.[3]

Especially troubling is the way the status of women as second-class citizens is being cemented in legislation. In Iraq, a law is being proposed that lowers to nine the legal age at which a girl can be forced into marriage. That same law would give a husband the right to deny his wife permission to leave the house. In Tunisia, your worries are about the imposition of sharia. In Afghanistan and Pakistan, by contrast, you have to fear being gunned down for the crime of attending school. And for young girls all over North Africa and beyond there remains the threat of female genital mutilation, a practice that certainly predates Islam but which is now almost entirely confined to Muslim communities. UNICEF estimates that more than 125 million women and girls have been cut in African and Arab nations, many of them majority Muslim.[4] As is gradually becoming clear, the practice is also widespread in immigrant communities in Europe and North America.

In the Islamic world, too many basic rights are circumscribed, and not only women's rights. Homosexuality is not tolerated. Other religions are not tolerated. Above all, free speech on the subject of Islam is not tolerated. As I know only too well, freethinkers who wish to question works such as the Qur'an or the hadith risk death.

Islam has had schism; it has never had Reformation. Early disputes in Islam produced fierce sectarianism that often involved bloodshed, but largely over technical questions. The biggest was about who should succeed the Prophet as leader of the *ummah*: the Sunnis wanted to select a caliph (literally a deputy) on the basis of merit, while the Shia insisted on an imam who was a relative of the Prophet. A smaller division was sparked by the question of whether Allah spoke in dictating the Qur'an. (One school of Islamic thought, the Mu'tazilite, argued that Allah does not have a human larynx and that the Qur'an is therefore not Allah's "speech.")[5]

The idea of "reform" in Islam has largely centered on the resolution of such narrow questions. Indeed, the term "*ijtihad*," the nearest thing to reform in Arabic, means trying to determine God's will on some new issue, such as: Should a Muslim pray on an airplane (a new technological invention) and, if so, how can he be sure he is facing Mecca? But the larger idea of "reform," in the sense of fundamentally calling into question central tenets of Islamic doctrine, has been conspicuous by its absence. Islam even has its own pejorative term for theological trouble-makers: "those who indulge in innovations and follow their passions" (the Arabic words *ahl al-bida, wa-l-ahwa'*).[6]

Tolerating Intolerance

Most Americans, and indeed most Europeans, would much rather ignore the fundamental conflict between Islam and their own worldview. This is partly because they generally assume that "religion," however defined, is a force for good and that any set of religious beliefs should be considered acceptable in a tolerant society. I can sympathize with that. In many respects, despite its high aims and ideals, America has found it difficult to make religious and racial tolerance a reality.

But that does not mean we should be blind to the potential consequences of accommodating beliefs that are openly hostile to Western laws, traditions, and values. For it is not simply a religion we have to deal with. It is a political religion many of whose fundamental tenets are irreconcilably inimical to our way of life. We need to insist that it is not we in the West who must accommodate ourselves to Muslim sensitivities; it is Muslims who must accommodate themselves to Western liberal ideals.

Unfortunately, not everyone gets this.

In the fall of 2014, Bill Maher, host of the HBO show *Real Time with Bill Maher*, held a discussion about Islam that featured the best-selling author Sam Harris, the actor Ben Affleck, and the *New York Times* columnist Nicholas Kristof. Harris and Maher raised the question of whether or not Western liberals were abandoning their principles by not confronting Islam about its treatment of women, promotion of jihad, and sharia-based punishments of stoning and death to apostates. To Affleck, this smacked of Islamophobia and he responded with an outburst of moralistic indignation. To applause from the audience, he heatedly accused Harris and Maher of being "gross" and "racist" and saying things no different from "saying 'you're a shifty Jew.'" Siding with Affleck, Kristof interjected that brave Muslims were risking their lives to promote human rights in the Muslim world.

After the show, during a discussion in the greenroom, Sam Harris asked both Ben Affleck and Nick Kristof, "What do you think would happen if we had burned a copy of the Qur'an on tonight's show?" Sam then answered his own question, "There would be riots in scores of countries. Embassies would fall. In response to our mistreating *a book*, millions of Muslims would take to the streets, and we would spend the rest of our lives fending off credible threats of murder. But when IS crucifies people, buries children alive, and rapes and tortures women by the thousands—*all in the name of Islam*—the response is a few small demonstrations in Europe and a hashtag [#NotInOurName]."

Shortly after the show was broadcast, a Pakistani-Canadian Muslim woman (and gay rights activist) named Eiynah wrote an open letter to Ben Affleck that summed up my feelings precisely:

> Why are Muslims being "preserved" in some time capsule of centuries gone by? Why is it okay that we continue to live in a world where our women are compared

to candy waiting to be consumed? Why is it okay for women of the rest of the world to fight for freedom and equality while we are told to cover our shameful bodies? Can't you see that we are being held back from joining this elite club known as the 21st century?

Noble liberals like yourself always stand up for the misrepresented Muslims and stand against the Islamophobes, which is great but who stands in my corner and for the others who feel oppressed by the religion? Every time we raise our voices, one of us is killed or threatened.

. . . What you did by screaming "racist!" was shut down a conversation that many of us have been waiting to have. You helped those who wish to deny there are issues, deny them.

What is so wrong with wanting to step into the current century? There should be no shame. There is no denying that violence, misogyny and homophobia exist in all religious texts, but Islam is the only religion that is adhered to so literally, to this day.

In your culture you have the luxury of calling such literalists "crazies." . . . In my culture, such values are upheld by more people than we realise. Many will try to deny it, but please hear me when I say that these are not fringe values. It is apparent in the lacking numbers of Muslims willing to speak out against the archaic Shariah law. The punishment for blasphemy and apostasy, etc, are tools of oppression. Why are they not addressed even by the peaceful folk who aren't fanatical, who just want to have some sandwiches and pray five times a day? Where are the Muslim protestors against blasphemy laws/apostasy? Where are the Muslims who take a stand against harsh interpretation of Shariah?[7]

Anyone for Apartheid?

One of the early suffragettes, Alva Belmont, said that American women must serve as a beacon of light, telling not only the story of what they have accomplished, but also representing a lasting determination that women around the world shall be "free citizens, recognized as the equals of men." Too often, when it comes to women's rights (and human rights more generally) in the Muslim world, leading thinkers and opinion makers have, at best, gone dark.

I cannot help contrasting this silence with the campaign to end apartheid, which united whites and blacks alike all over the world beginning in the 1960s. When the West finally stood up to the horrors of South African apartheid, it did so across a broad front. The campaign against apartheid reached down into classrooms and even sports stadiums; churches and synagogues stood united against it across the religious spectrum. South African sports teams were shunned, economic sanctions were imposed, and intense international pressure was brought to bear on the country to change its social and political system. American university students erected shantytowns on their campuses to symbolize their solidarity with those black South Africans confined to a life of degradation and impoverishment inside townships.

Today, with radical Islam, we have a new and even more violent system of apartheid, where people are targeted not for their skin color but for their gender, their sexual orientation, their religion, or, among Muslims, the form of their personal faith.

I have spent more than a decade fighting for women's and girls' basic rights. I have never been afraid to ask difficult questions about the role of religion in that fight. As I have repeatedly said, the connection between violence and Islam is too clear to be ignored. We do no favors to Muslims when we shut our eyes to this link, when we excuse rather than reflect. We need to ask: Is the concept of holy war compatible with our ideal of religious toleration? Should it be blasphemy—punishable by death—to question the applicability of certain seventh-century doctrines to our own era? Why, when I have made these arguments, have I received so little support and so much opprobrium from the very people in the West who call themselves feminists, who call themselves liberals?

I do not expect our political leadership to take the lead in directly challenging the inequities of political Islam. The ideological self-confidence that characterized Western leaders during the Cold War has given way to a feeble relativism. Instead, this campaign for female, gay, and minority rights needs to come from elsewhere: from the men who built Silicon Valley's social networks, whose instincts are deeply libertarian; from our entertainment capital, Hollywood, where at least the old hands still remember the era of blacklists and witch hunts; from our civil society, from human rights activists, from feminists, and from lesbian, bisexual, gay, and transgender communities; as well as from organizations like the ACLU who, if they still stand for anything, can hardly ignore the way civil liberties are being trampled all over the Muslim world. They must remember Alva Belmont's words. They must light their beacons.

A Unique Role for the West

Whenever I make the case for reform in the Muslim world, someone invariably says: "That is not our project—it is for Muslims only. We should stay out of it." But I am not talking about the kind of military intervention that has got the West into so much trouble over the years.

For years, we have spent trillions on waging wars against "terror" and "extremism" that would have been much better spent protecting Muslim dissidents and giving them the necessary platforms and resources to counter that vast network of Islamic centers, madrassas, and mosques which has been largely responsible for spreading the most noxious forms of Islamic fundamentalism. For years, we have treated the people financing that vast network—the Saudis, the Qataris, and the now repentant Emiratis—as our allies. In the midst of all our efforts at policing, surveillance, and even military action, we in the West have not bothered to develop an effective counternarrative because from the outset we have denied that Islamic extremism is in any way related to Islam. We persist in focusing on the violence and not on the ideas that give rise to it.

Yet here is another conflict that we can take inspiration from as we embark on this process: the Cold War.

Islam is not communism, of course, but in certain respects it is just as contemptuous of human rights, and Islamic republics have proved almost as brutal toward their own citizens as Soviet republics once were. Yet we have welcomed fundamentalist preachers into our cities and have stood idly by as thousands of disaffected young people have been radicalized by their rantings. Worse, we have made almost no attempt to counter the proselytizing of the Medina Muslims. If we continue this policy of nonintervention in the culture war, we will never extricate ourselves from the actual battlefield. For we cannot fight an ideology solely with air strikes and drones or even boots on the ground. We need to fight it with ideas—with better ideas, with positive ideas. We need to fight it with an alternative vision, as we did in the Cold War.

The West did not win the Cold War simply through economic pressure or building new weapons systems. From the beginning, the United States recognized that this was also going to be an intellectual contest. Aside from a few "useful idiots" on leftist campuses, we did not say the Soviet system was morally equivalent to ours; nor did we proclaim that Soviet communism was an ideology of peace.

Instead, through a host of cultural initiatives funded directly or indirectly by the CIA, the United States encouraged anti-Communist intellectuals to counter the influence of Marxists and other fellow-travelers of the Left. The Congress for Cultural Freedom, dedicated to defending the non-Communist Left in the battle of ideas in the world, opened in Berlin on June 26, 1950. Leading intellectuals such as Bertrand Russell, Karl Jaspers, and Jacques Maritain agreed to serve as honorary chairmen. Many of the members were former Communists such as Arthur Koestler who warned against the dangers of totalitarianism on the basis of personal experience.[8] Magazines such as *Encounter* (UK), *Preuves* (France), *Der Monat* (Germany), and *Quadrant* (Australia) were made beneficiaries of American support.[9] The Free Europe Press mailed numerous books to dissidents in Eastern Europe, sneaking their materials past the censors wherever they could. By the end of the

Cold War, "it was estimated that over ten million Western books and magazines had infiltrated the Communist half of Europe through the book-mailing program."[10]

How much did these efforts cost? In the case of the Congress for Cultural Freedom, surprisingly little. In 1951, the budget of the Congress for Cultural Freedom seems to have been about $200,000, or approximately $1.8 million in 2014 dollars.[11] Contrast the small budget of the Congress for Cultural Freedom with the enormous sums the United States has spent since 2001 against what policymakers call "terror" or "extremism." A 2013 analysis of the so-called black budget suggested that the United States has spent more than $500 *billion* on various intelligence agencies and efforts from 2001 to 2013.[12] The economist Joseph Stiglitz has calculated the cost of the military intervention in Iraq to be between $3 and $5 trillion.[13]

This strategy is unsustainable. For one, the United States cannot afford to continue fighting a war of ideas solely by military means. Second, by ignoring the ideas that give rise to Islamist violence we continue to ignore the root of the problem.

Instead, modeled on the cultural campaigns of the Cold War, there must be a concerted effort to turn people away from fundamentalist Islam. Imagine a platform for Muslim dissidents that communicated their message through YouTube, Twitter, Facebook, and Instagram. Imagine ten reformist magazines for every one issue of IS's *Dibaq* or Al-Qaeda's *Inspire*. Such a strategy would also give us an opportunity to shift our alliances to those Muslim individuals and groups who actually share our values and practices—those who fight for a true Reformation and who find themselves maligned and marginalized by those nations and leaders and imams whom we now embrace as allies.

In the Cold War, the West celebrated dissidents such as Aleksandr Solzhenitsyn, Andrei Sakharov, and Václav Havel, who had the courage to challenge the Soviet system from within. Today, there are many dissidents who challenge Islam—former Muslims, and reformers—but the West either ignores them or dismisses them as "not representative." This is a grave mistake. Reformers such as Tawfiq Hamid, Irshad Manji, Asra Nomani, Maajid Nawaz, Zuhdi Jasser, Saleem Ahmed, Yunis Qandil, Seyran Ateş, Bassam Tibi, and many others must be supported and protected. They should be as well known as Solzhenitsyn, Sakharov, and Havel were in the 1980s—and as well known as Locke and Voltaire were in their day, when the West needed freethinkers of its own.

CONCLUSION

THE MUSLIM REFORMATION

Today there is a war within Islam—a war between those who wish to reform (the Modifying Muslims or the dissidents) and those who wish to turn back to the time of the Prophet (the Medina Muslims). The prize over which they fight is the hearts and minds of the largely passive Mecca Muslims.

For the moment, measured by four yardsticks, the Medina side seems to be winning. One is the scale of individuals leaving the Mecca side and joining the Medina side (what in the West we call "radicalization"). The second metric is attention: the Medina Muslims attract media attention through statements and acts of violence that shock the world. The third metric is resources: through *zakat* (charity), crime, the violent seizing of territory and property, support from rogue states, and petrodollars, Medina Muslims have vast resources. The

Modifying Muslims have virtually none. Pushed to make a choice between earning a living and campaigning for religious reform, most Modifiers soon opt for the former. The fourth metric is one of coherence. In many ways this is the most important advantage the Medina Muslims have over the Modifier Muslims. The latter are faced with the daunting—and dangerous—task of questioning the fundamentals of their faith. All the Medina Muslims have to do is pose as its defenders.

Yet I believe a Muslim Reformation is coming. In fact, it may already be here. I think it is plausible that the Internet will be for the Islamic world in the twenty-first century what the printing press was for Christendom in the sixteenth. I think it is plausible that the violent Islamists I have called the Medina Muslims are the modern counterparts to the millenarian sects of pre-Reformation Europe and that a quite different reform movement is already taking shape in the cities of the Middle East and North Africa. Above all, I believe that the upsurge of popular protest that we call the Arab Spring contained within it some of the seeds of a true Muslim Reformation, despite the obvious and predictable failure of the political revolution to live up to Western hopes of a Middle Eastern 1989.

Much at this early stage is uncertain. The only real certainty about the Muslim Reformation is that it will not look much like the Christian one. There are such fundamental differences between the teachings of Jesus and Muhammad, to say nothing of the radically different organizational structures of the two religions—the one hierarchical and distinct from the state, the other decentralized yet aspiring to political power—that any analogy is bound to break down.

When I first conceived of writing a book about a Reformation of Islam, I imagined it as a novel. Entitled *The Reformer*, it was going to tell the story of a charismatic young imam in London who would emerge as a modern-day Muslim Luther. I abandoned the idea because such a book was bound to be dismissed as fanciful.

The Muslim Reformation is not fiction. It is fact. Over the past few years, dozens if not hundreds of developments have convinced me that, while Islam's problems are indeed deep and structural, Muslim people are like everyone else in one important respect: most want a better life for themselves and their children. And increasingly they have good reasons to doubt that the Medina Muslims can deliver it.

It is no accident that some of the most vocal critics of Islam today are, like me, women. For there is no more obvious incompatibility between Islam and modernity than the subordinate role assigned to women in sharia law. That subordinate role has long been the justification for a litany of abuses of women in the Muslim world, such as male guardianship, child marriage, and marital rape. Just as the surge of sexual assaults was one of the most disturbing features of the Egyptian Revolution, so the response of groups like Tahrir Bodyguard and Operation Anti-Sexual Harassment was one of the most heartening. We are seeing similar movements in Lebanon and Jordan, notably the protest against Article 308, the Jordanian law that allows rapists to marry their victims to avoid going to jail. Iran is an especially interesting case, for there thirty years of Islamist rule appear to have failed to prevent a significant shift in attitudes toward female sexuality.

Yet it would be a mistake to think of this movement in narrowly feminist terms. Although it is women who are spearheading change, there are other issues in play besides the status of women as second-class citizens. In some parts of Africa, we are seeing waves of conversion from Islam to

Christianity. Another pioneer of change is Walid Husayin, the Palestinian skeptic jailed for antireligious agitation. Then there are the Muslims who speak out for toleration, such as the Turkish columnist and TV commentator Aylin Kocaman, who has defended Israel and rejected Islamist calls for violence against Jews, or Nabil al-Hudair, an Iraqi Muslim who has spoken up for the rights of his Jewish fellow countrymen.

There really are tides in the affairs of men—and women, too. I believe this is one of those historic tides.

Why the Tide Is Turning

Three factors are combining today to enable real religious reform:

- The impact of new information technology in creating an unprecedented communication network across the Muslim world.
- The fundamental inability of Islamists to deliver when they come to power and the impact of Western norms on Muslim immigrants are creating a new and growing constituency for a Muslim Reformation.
- The emergence of a political constituency for religious reform emerging in key Middle Eastern states.

Together, I believe these three things will ultimately turn the tide against the Islamists, whose goal is, after all, a return to the time of the Prophet—a venture as foredoomed to failure as all attempts to reverse the direction of time's arrow.

As we have seen, technology is empowering not only the

jihadists. It is also empowering those who would oppose them in the name of human rights for all, regardless of religion. (Without the assistance of Google, for example, it would have been far harder for me to write this book.) In November 2014, an Egyptian doctor coined an Arabic hashtag that translates as "why we reject implementing sharia"; it was used five thousand times in the space of twenty-four hours, mostly by Saudis and Egyptians. In language that would have been unthinkable just a few years ago, a young Moroccan named Brother Rachid last year called out President Obama on YouTube for claiming that Islamic State was "not Islamic":

> Mr President, I must tell you that you are wrong about ISIL. You said ISIL speaks for no religion. I am a former Muslim. My dad is an imam. I have spent more than 20 years studying Islam. . . . I can tell you with confidence that ISIL speaks for Islam. . . . ISIL's 10,000 members are all Muslims. . . . They come from different countries and have one common denominator: Islam. They are following Islam's Prophet Muhammad in every detail. . . . They have called for a caliphate, which is a central doctrine in Sunni Islam.
>
> I ask you, Mr. President, to stop being politically correct—to call things by their names. ISIL, Al Qaeda, Boko Haram, Al-Shabaab in Somalia, the Taliban, and their sister brand names, are all made in Islam. Unless the Muslim world deals with Islam and separates religion from state, we will never end this cycle. . . . If Islam is not the problem, then why is it there are millions of Christians in the Middle East and yet none of them has

> ever blown up himself to become a martyr, even though they live under the same economic and political circumstances and even worse? . . . Mr. President, if you really want to fight terrorism, then fight it at the roots. How many Saudi sheikhs are preaching hatred? How many Islamic channels are indoctrinating people and teaching them violence from the Quran and the hadith? . . . How many Islamic schools are producing generations of teachers and students who believe in jihad and martyrdom and fighting the infidels?[1]

(Having been saying such things for more than thirteen years, I feel a surge of hope when I read words like those in *The New York Times*.)

Brother Rachid is a Moroccan convert to Christianity who broadcasts on a television station, Al-Hayat, based in Egypt. His story perfectly illustrates how fast things are changing in North Africa and the Middle East. Religious minorities, as well as women and gay people, remain highly vulnerable in the Middle East and North Africa. But precisely because of their sufferings, I think it is ever more likely that they will ultimately unite against Islam's religious apartheid. When I see millions of women in Afghanistan defying threats from the Taliban and lining up to vote; when I see women in Saudi Arabia defying an absurd ban on female driving; and when I see Tunisian women celebrating the conviction of a group of policemen for a heinous gang rape, I feel more optimistic than I did a few years ago.

In short, I agree with Malala Yousafzai, the Nobel Peace Prize–winning Pakistani schoolgirl whom the Taliban tried to kill:

> The extremists are afraid of books and pens. The power of education frightens them. They are afraid of women. The power of the voice of women frightens them. That is why they are blasting schools every day—because they were and they are afraid of change, afraid of the equality that we will bring to our society. They think that God is a tiny, little conservative being who would send girls to the hell just because of going to school.[2]

Here, surely, is the authentic voice of a Muslim Reformation.

Change is also under way in the Muslim communities of the Western world. True, further Muslim immigration to Europe and North America will very likely increase the tensions between Westerners and Muslims. Yet even as the probability of such conflict increases, so too does the exposure of second- and third-generation Muslims to Western values and freedoms. Yes, some will withdraw into a cocoon of denial, and others will become Medina Muslims in reaction against the dissonances they experience. In the long run, however, these options are far less appealing than the third option of religious reform.

Finally, there is the horrified reaction of many Muslims to the atrocities committed by Al-Qaeda, IS, and Boko Haram, which has led some Muslim political leaders to get serious about taking Islam back from the extremists. The government of the United Arab Emirates has called the threat posed by "Islamic extremism" a "transnational cancer" requiring an "urgent, coordinated and sustained international effort to confront" it.[3] The fight against radical Islam, the UAE ambassador to the United States insisted, "must be waged not only on the battlefield but also against the entire militant *ideological* and financial complex that is the lifeblood of extremism."[4] Before an audience of Muslim clergy, as we have seen, the president of Egypt himself has called for a "religious revolution." That is the kind of support a Reformation cannot do without if it is to succeed.

The fact that President Sisi elected to make his call for religious revolution at Al-Azhar—the preeminent institution of Sunni religious learning in the world—was highly significant. For Al-Azhar has long been the citadel of clerical conservatism, ruthlessly resisting even the discussion of meaningful reforms to Islam.[5] In June 1992, for example, an Egyptian academic and human rights activist named Farag Foda was shot dead as he left his office. For years, Foda had defended secular policies and criticized sharia law, arguing for a separation of religion and politics. Two weeks before Foda's death, the widely respected cleric Muhammad al-Ghazali, a senior figure at Al-Azhar, had declared Foda to be an apostate, knowing full well that under Islamic religious law, the punishment for apostasy is death.[6] Activists of the Islamic group Gama'a al Islamiyya subsequently killed Foda, heavily injuring bystanders (including Foda's son) in the process. "Al-Azhar issued the sentence and we carried out the execution," the group stated.[7] Al-Ghazali, the cleric who had declared Foda an apostate, subsequently testified on behalf of Foda's killers, arguing that the presence of an apostate inside the community constituted a threat to the nation.[8] Though now deceased, al-Ghazali remains a venerated figure among Islamic scholars,[9] while Al-Azhar as an institution has never expressed any contrition for its role in Foda's death.

It is precisely institutions like Al-Azhar that stand in the way of a Muslim Reformation. If the Egyptian government is prepared to take on Al-Azhar, the times are indeed changing.

Je Suis Charlie

There is one final reason I am optimistic. I begin to hope that the West may finally be coming to its senses.

Over the past twenty years, terrified of appearing culturally insensitive or even racist, Western nations have bent over backward to accommodate the demands of their Muslim citizens for special treatment. We appeased the Muslim heads of government who lobbied us to censor our press, our universities, our history books, our school curricula. We appeased leaders of Muslim organizations in our societies, who asked universities to disinvite speakers deemed "offensive" to Muslims. Instead of embracing Muslim dissidents, Western governments treated them as troublemakers and instead partnered with all the wrong people—groups such as the Council on American-Islamic Relations.[10] And we even subsidized the jihadists. (For example, the man who killed Theo van Gogh was living off Dutch welfare benefits.)

Yet I dare to hope that what happened in Paris in January 2015 may prove to be a turning point. It was not that the *Charlie Hebdo* massacre was especially bloody. Many more people had died in the Taliban attack on the Army Public School in Peshawar, Pakistan, in December 2014. Many more people died in the Boko Haram attack on Baga in Nigeria in the same week as the attack in Paris. Rather, it was the fact that more than a dozen people were murdered because they had drawn and published cartoons of the Prophet Muhammad.

There were, of course, the usual craven editorials and press statements by moral idiots arguing that the editors of the magazine had lacked "common sense" in offending Muslims, and that nevertheless the violence had nothing to do with Islam. But for the millions of people who took to the streets bearing "Je Suis Charlie" signs, these arguments clearly were not reassuring.

As of this writing, ten thousand military and security personnel have been deployed across France as authorities brace for more attacks. Even to me, just a week ago, such a militarization of policing in one of the West's largest and oldest democracies would have been unthinkable. France's prime minister, Manuel Valls, said three days after the attack that France was at war with "radical Islam." The French, once so critical of the United States after 9/11 (not least for the sweeping scope of the Patriot Act), are now following in the footsteps of George W. Bush. Stephen Harper, the prime minister of the other great French-speaking democracy, Canada, explicitly connected the *Charlie Hebdo* attack to the "international jihadist movement." "They have declared war on anybody who does not think and act exactly as they wish they would think and act," Harper said. "They have declared war and are already executing it on a massive scale on a whole range of countries with which they are in contact, and they have declared war on any country, like ourselves, that values freedom, openness and tolerance. We may not like this and wish it would go away, but it is not going to go away."

At a time like this, the claims that the "extremists" have nothing to do with the "religion of peace" simply cease to be credible. The enemy in this war is saying just the opposite. Consider, for example, the book written by the Al-Qaeda operative Abu Musab al-Suri, entitled *The Call to Global Islamic Resistance.* As the enemies of Islam, al-Suri lists: the Jews, America, Israel, the Freemasons, the Christians, the Hindus, apostates (including established Muslim leaders, officials, and their security apparatus), hypocritical scholars, educational systems, satellite TV channels, sports, and all arts and entertainment venues.[11] This would be comical if it were not so deadly serious.

Western leaders who insist on ignoring such explicit threats run two risks. Not only do their words ("Islam belongs to Germany") embolden the zealots. They also create a political vacancy. Even before *Charlie Hebdo,* Germans were protesting under the banner of Pegida (short for "Patriotic Europeans Against the Islamization of the West") in Dresden, Berlin, Munich, and Leipzig. All over Europe, populist parties are mobilizing voters in increasing numbers against immigration and Islam, from the National Front in France to the Sweden Democrats. It can be in nobody's interests for Europe to slide in this way down a perilous path of polarization.

Instead, as briefly happened in Paris in the days after the massacre, we in the West need to unite. But we need to be clear about what we are uniting for, and what we are uniting against.

In all holy books, in the Bible as well as the Qur'an, you will find passages that sanction intolerance and inequity. But in the case of Christianity, there was change. In that process of change, the people who wanted to uphold the status quo made the same arguments that present-day Muslims are giving: that they were offended, that the new thinking was blasphemy. In effect, it was *through* a process of repeated blasphemy that Christians and Jews evolved and grew into modernity. That is what art did. That is what science did. And yes, that is what irreverent satire did.

The Muslim Reformation is not going to come from Al-Azhar. It is more likely to come from a relentless campaign of blasphemy. So when a Muslim sees you reading this book and says, "I am offended, my feelings are hurt," your reply should be: "What matters more? Your sacred text? Or the life of this book's author? Your sacred text? Or the rule of law? Human life, human freedom, human dignity—they all matter more than any sacred text." Christians have been through this, Jews have been through it. It's now time for Muslims to go through it. In that sense—in the sense that I passionately believe in the world-changing power of blasphemy—*je suis Charlie.*

Yet we need to do more than merely blaspheme. We need to reform.

The Five Amendments, Restated

The tenth- and eleventh-century Islamic legal scholar al-Mawardi, writing in *The Ordinances of Government*, says: "If an innovator appears or a holder of suspect views goes astray, the imam should explain and clarify the correct view to him, and make him undergo the penalties appropriate to him, so that the religion may be preserved from flaws and the community preserved from error."[12] I know that anyone who advocates reforming Islam runs a risk. So let me be unambiguously clear. I am not advocating a war—quite the contrary. I am explicitly arguing for peaceful reform: for a cultural campaign aimed at doctrinal change.

As I have argued, there are five core concepts in Islam that are fundamentally incompatible with modernity:

1. the status of the Qur'an as the last and immutable word of God and the infallibility of Muhammad as the last divinely inspired messenger;
2. Islam's emphasis on the afterlife over the here-and-now;
3. the claims of sharia to be a comprehensive system of law governing both the spiritual and temporal realms;
4. the obligation on ordinary Muslims to command right and forbid wrong;
5. the concept of jihad, or holy war.

My "five theses" are simply that these concepts must be amended in ways that make being a Muslim more readily compatible with the twenty-first-century world. Muslim clerics need to acknowledge that the Qur'an is not the ultimate repository of revealed truth. They need to make explicit that what we do in this life is more important than anything that could conceivably happen to us after we die. It is just a book. They need to make clear that sharia law occupies a circumscribed role and is clearly subordinate to the laws of the nation-states where Muslims live. They need to put an end to the practice of delegated coercion that inflicts conformity at the expense of creativity. And they need to disavow completely the concept of jihad as a literal call to arms against non-Muslims and those Muslims they deem apostates or heretics.

This Reformation would not only benefit women, gays, and religious minorities. I believe it is also in the interests of Islam itself. In order to avoid eventual collapse, even the most revered structure requires renovation. Mere restoration is no longer a plausible option for Islam, no matter how much blood the Islamists shed. Indeed, the more blood they shed, the more they risk bringing the entire structure crashing down upon their heads.

How long will the rest of us have to wait for this Reformation to succeed in transforming Islam as deeply as the original Reformation transformed Christianity? In the last decade, many thousands of innocent people have lost their lives in an escalating sectarian conflict that rages across borders. Tens of millions of decent men and women and their children remain trapped within failing states, stagnating economies, and repressive societies. Will the Muslim Reformation be widespread or localized (after all, the Protestant Reformation did not succeed in all of Christendom)? Will the Muslim Reformation produce wars of religion, like its Christian predecessor, before its more beneficial effects make themselves felt?

The answers to these questions depend above all on Muslims and the choices they make. But they also depend to some extent on the choices we in the West make. Do we help the Reformation? Or do we unwittingly undermine it?

It will not be easy to bring about this change. But perhaps the words of two thinkers, one an Islamic heretic and one a master of the Western Enlightenment, can give us encouragement.

In 1057, the Syrian poet and philosopher Abul 'Ala al-Ma'arri died. In his lifetime, for the act of forgoing meat and being a vegetarian, he was branded a heretic. He was also branded a heretic for his poetry and other fictional writings, including *The Epistle of Forgiveness*, in which he imagined a journey to heaven and to hell.[13]

Although he is largely unknown in the West, his work is regarded as a forerunner of Dante's *Divine Comedy*, and over the years, statues of him have been erected around his home region, south of Aleppo. In 2013, jihadists, primarily with the Al-Nusra Front, began attacking and beheading his statues. There are multiple theories about the attacks, including one that perhaps al-Ma'arri is related to President Assad. But the more plausible explanation is that nothing—not even the passage of a thousand years—can expunge the guilt of the heretic. The stigma of heresy is eternal.[14]

And what did al-Ma'arri write that was so heretical? Here are a few of his lines: "Shall I go forth from underneath this sky? How shall I escape? Whither shall I flee?" And: "God curse people who call me an infidel when I tell them the truth!" And: "I lift my voice whene'er I talk in vain, / But do I speak the truth, hushed are my lips again."[15]

I find those lines almost unbearably moving. And yet, nearly a thousand years after they were written, I am certain that the time for heretics to speak the truth with impunity has at last arrived. And for those still unsure how they should react to the words of a heretic, I turn again to Voltaire, the freest of freethinkers. "I disapprove of what you say," he is said to have written to Claude Helvétius, "but I will defend to the death your right to say it."

The dawn of a Muslim Reformation is the right moment to remind ourselves that the right to think, to speak, and to write in freedom and without fear is ultimately a more sacred thing than any religion.

· APPENDIX ·

Muslim Dissidents and Reformers

The best evidence that a Muslim Reformation is actually under way is the growing number of active dissidents and reformers around the world. It would be quite wrong of me to publish this book without acknowledging them and their often courageous contributions. Broadly speaking, they can be grouped into three broad categories: dissidents in the West, dissidents in the Islamic world, and clerical reformers.

Dissidents in the West

There is a growing number of ordinary Muslim citizens in the West who are currently braving death threats and even official punishment in dissenting from Islamic orthodoxy and calling for the reform of Islam. These individuals are not clergymen but "ordinary" Muslims, generally educated, well read, and preoccupied with the crisis of Islam.

Among them are Maajid Nawaz (UK), Samia Labidi (France), Afshin Ellian (Netherlands), Ehsan Jami (Netherlands), Naser Khader (Denmark), Seyran Ateş (Germany), Yunis Qandil (Germany), Bassam Tibi (Germany), Raheel Raza (Canada), Zuhdi Jasser (U.S.), Saleem Ahmed (U.S.), Nonie Darwish (U.S.), Wafa Sultan (U.S.), Saleem Ahmed (U.S.), Ibn Warraq (U.S.), Asra Nomani (U.S.), and Irshad Manji (U.S.).

These individuals are not clerics, but informed citizens speaking out on the basis of reason and conscience. They are urging either a fundamental reinterpretation of Islam or a change in the core doctrines of Islam. Some of them have left the faith, seeking reform from the outside, whereas others seek to reform Islam from within.[1] Their arguments focus on the importance of viewing the Qur'an and the hadith in a historical context and on respecting man-made civil laws as legitimate, overriding sharia religious law.

Zuhdi Jasser, an American Muslim physician, is the founder of the American Islamic Forum for Democracy based in Phoenix, Arizona. Jasser has embarked on the "Jefferson project" for Islam. He favors the separation of mosque and state, which will "include the abrogation of all blasphemy and apostasy laws" currently used to stifle Muslim reformers. His aim is to reform Islam and place civil law above sharia law:

> If government enacts the literal laws of God rather than natural law or human law, then government becomes God, and abrogates religion and the personal nature of the relationship with God. Governmental law should be based on and debated in reason, not from scriptural exegesis.[2]

Saleem Ahmed, a Muslim now living in Hawaii, was born in India and raised in Pakistan. Ahmed founded the Honolulu-based

All Believers Network in 2003, promoting genuine interfaith dialogue. Its board has individuals from numerous religions, including Buddhism, Christianity, Taoism, and Islam. Ahmed argues that the more political and violent verses of the Qur'an are superseded by spiritual passages having universal applicability.[3] He has written a book arguing for a fundamental reform of Islamic doctrine. A number of fellow Muslims have called Ahmed a *kafir* (nonbeliever) and his local imam has criticized him for "diluting our religion."[4] Ahmed says that his role model is Gandhi.

Yunis Qandil, now living in Germany, was born in Amman, Jordan. He is the son of Palestinian refugees. In his later youth he became closely involved in a Salafi mosque for five years before turning to the Muslim Brotherhood for another four years. He moved to Germany in 1995 and increasingly "sought to combine his spirituality with a secular stance regarding politics."[5] Qandil is critical of groups such as the Muslim Brotherhood that seek to create a "parallel society" of European Muslims, preventing individual Muslims from fully integrating into their host societies.[6] Even if Islamists such as the Muslim Brotherhood oppose the use of violence in the short term, they are not true partners for genuine integration and peaceful coexistence in a pluralist democracy. Qandil continues his work for the separation of mosque and state.

Samia Labidi, now living in France, was born in Tunisia in 1964. She attended an Islamic school and grew up in a traditional but tolerant family.[7] When she was eleven, her sister married one of the founders of the Islamist group MTI, known as El Nahda (the Renaissance). Her family then became Medina Muslims and Labidi began wearing the veil.[8] Labidi's mother found the situation too confining and left Tunisia to live with her brother in France. Labidi, too, felt that she could barely breathe:

> My mind was sterilized gradually, unable to have access to freedom of thought, to myself. . . . Women continued to be treated like incapable beings who need to be systematically under the guardianship of a close male relative in order to move, to exist, or even to breathe.[9]

When she was eighteen, Labidi left Tunisia and went to Paris, earning a master's degree in philosophy from Université de Paris X Nanterre. Labidi's brother, meanwhile, became radicalized before abjuring terrorism. Labidi has written about her brother's radicalization[10] and now argues for reforming Islam: "Ultimately," she writes, "the solution lies in separating religion from politics, particularly in that part of the globe that is still suffering from this amalgam between . . . temporal . . . and spiritual power."[11] Labidi remains highly active in groups that are seeking to give secular French Muslims a voice.[12]

Seyran Ateş is a German lawyer of Turkish descent. Ateş moved with her family from Turkey to Germany as a six-year-old in 1969. Just before she turned eighteen, she left her parents' home, moved in with a German man, and studied law.[13] As an attorney specializing in family law, Ateş represented numerous Muslim women for two decades in cases involving abusive marriages, forced marriages, and divorce proceedings.

Through her work, Ateş has seen the dark side of excessively tolerant multiculturalism. According to Ateş, forced marriages are locking up German-born Muslims in separate Islamic enclaves to the point that tens of thousands of women are so isolated from German society that they are unable even to call an ambulance. There has been excessive tolerance for the repressive side of Islam, something Ateş calls the "multicultural mistake," the title of one of several books she has written.

Before she was pressured to stop her public appearances by

security threats, Ateş argued that Islam needs "a sexual revolution" to emancipate women as equals: "Part of the process is that sexuality [in Islam] has to be recognized as something that every individual determines for himself or herself."[14] She has proposed creating a mosque that would welcome Sunnis and Shiites and treat men and women equally, allowing men and women to pray together and women to serve as imams in mixed congregations.

Ateş argues that Islam must be completely separated from politics: "If we are going to stop that movement and separate politics from religion," Ateş says, "then we will have chance for Islam to be compatible with democracy."[15]

Citizen Reformers in the Islamic World

In the Islamic world, too, a growing number of ordinary citizens are calling for reform. These voices include the Egyptian Kareem Amer, the Palestinian Walid Husayin, the Turk Aylin Kocaman, the Iraqi Nabil al-Haidari, the Pakistani Luavut Zahid, the Saudi Arabians Hamza Kashgari and Raif Badawi, and the Bangladeshi Taslima Nasrin.

Kareem Amer (real name Abdel Suleiman) is an Egyptian and a former student at Al-Azhar. In 2005, after Muslims attacked a Coptic church, Amer called Muhammad and his seventh-century followers the *sahaba*—"spillers of blood"—for their teachings on warfare.[16] Amer criticized Al-Azhar as being a force for Islamic orthodoxy and intolerance of reformist views. Early in 2006, he was expelled for criticizing the extreme dogma of his Islamic instructors, writing on his blog that "professors and sheikhs at al-Azhar who . . . stand against anyone who thinks freely" would "end up in the dustbin of history."[17] Amer also criticized the autocratic rule of then-President Hosni Mubarak. He was sentenced to four years in prison in 2007 before being released in 2010 after being beaten in confinement. He exemplifies those young Egyptians who question not only political but also religious authoritarianism.

Walid Husayin, about thirty years old, is a Palestinian skeptic who has described the Islamic God as "a primitive, Bedouin and anthropomorphic God."[18] On Facebook, Husayin also satirized various Qur'anic verses. Husayin is in every sense an irreverent freethinker who in the West might have found work as a comedian or satirist. Many Palestinians, however, responded with anger to Husayin's criticism of Islam, accusing him of working for the Mossad, Israel's intelligence agency. Some residents of his hometown called on him to be killed "as a warning to others."[19] Husayin responded that his critics "actually don't get that people are free to think and believe in whatever suits them."[20] After being jailed for a month, and under heavy pressure, Husayin apologized.[21]

Luavut Zahid, a Pakistani writer and women's rights advocate, wrote in April 2014 that Muslims had to make some significant changes to their religion, and that the crisis of Islam could not be blamed on outsiders:

> The tactics of terror used by Islamic countries and Muslims at large in general ensure that people will either put up with them, or shut up and leave. There is no concept of freedom of speech, and there is furthermore no concept of criticism. . . . A more pertinent question instead would be why people never spring into action when someone passes a fatwa allowing and requiring female genital mutilation. If it is not real Islam to circumcise young girls, then why did people realise it only after [Ayaan] Hirsi Ali spoke about it? . . . Does she at times sound too extreme? Definitely.

> But stop for a second and ask yourself this: how many Muslims has she killed? How many Muslims have had to go into hiding because of her? The onus for change lies with Muslims alone. If they are so hell bent on proving that this extreme interpretation of their faith is wrong, then they need to come forward and start transforming things from the inside. Hirsi Ali cannot and should not be called an Islamophobe only because she loudly repeats the things that she has experienced, and continues to see happening around her, and all in the name of God.[22]

Taslima Nasrin, an apostate born in Bangladesh currently living in India, has said that "what is needed is a uniform civil code of laws that is not based on religious dogmas, and that is equally applicable to men and women."[23] The rule of civil law rather than sharia law will ensure all citizens are treated as equals, regardless of their private religious affiliation. This would entail a full separation of mosque and state.

Dissident Clerics

My own sense is that a Muslim Reformation will not come from within the ranks of the Islamic clergy. In the current crisis of Islam, however, there is a growing chorus of Muslim clerics calling for reform of existing Islamic doctrine. Such reformers can be found among both Sunni and Shia clerics, in the Islamic world as well as in the West. These clerics ought to be distinguished from what I would call "fake" reformers, who may condemn the violence used by Al-Qaeda and Islamic State while fervently working toward the imposition of sharia through nonviolent means. That is not what a real "reformer" is, though Western governments—

including the U.S. government—have often made the mistake of partnering with such individuals.[24] A *real* reformer is a cleric who not only rejects violence in the short term but *also* favors changing certain core religious doctrines of Islam.

These clerical reformers differ on the specific substance of reforms. Some (such as al-Ansari) favor reinterpretation of Islamic doctrine while respecting, for example, the integrity of the text of the Qur'an. Others (such as al-Qabbanji) view the Qur'an as a human-influenced text subject to far-reaching reinterpretation.

A description of some clerical reformers will reveal that there are meaningful efforts at present to reform Islam from within, though my own sense is that citizen-reformers will ultimately be more powerful than clerics in reforming Islam.

Imam Yassin Elforkani, a Sunni preaching in the Netherlands, has argued that "a new theology must arise in a Dutch context."[25] Though Elforkani views the Qur'an as a divine text (in that regard adhering to orthodoxy), he insists that "all interpretations of the Qur'an are the work of human beings" and subject to change. About young Dutch Muslims who leave the Netherlands to join IS, he says, "We [Muslims] can't permit ourselves to look away, we've got to think critically about ourselves. . . . These young people left with ideals that did not fall from the sky. Those ideals coincide with elaborate theories, with concepts from Islamic theology that have been taught for decades."[26]

Elforkani has expressed himself critically about the theory of the Caliphate and the activities of IS: "The concept of the Caliphate, of the global rule of Islam—sorry, but that is not of this era, is it? But if we do not develop alternatives to this, IS will only gain more and more ground." Elforkani has received numerous death threats in the Netherlands for explicitly calling for theological reforms within Islam.

In the Islamic world, a number of clerics are publicly calling for theological reforms within Islam. The Sunni 'Abd Al-Hamid al-Ansari is a former dean of Islamic law at Qatar University. Born in Doha in 1945, al-Ansari has defended liberal Muslims for years. Rejecting calls by Islamic preachers for young Muslims to love death, Ansari has said: "I would like the religious scholars, through their religious discourse, to make our youth love life, and not death."[27] Al-Ansari has called for a fundamental overhaul of educational systems in the Islamic world to encourage critical thinking. He has called for Arab freethinkers to be able to sue inflammatory Islamic preachers for harm that befalls them as a result of their sermons.[28]

Ahmad al-Qabbanji is a Shiite cleric who has proposed to change core aspects of Islam's doctrines. Al-Qabbanji was born in Najaf, Iraq, in 1958, and studied Islamic jurisprudence at the Shiite Hawza of his hometown in the 1970s. He has said openly:

> I have deviated from [t]his religion, every bit of which I reject. Let them say that I am an apostate and a heretic. It is true. I am an apostate from their religion, which stirs nothing but hatred of the other—a religion devoid of beauty, devoid of love, devoid of humanity.[29]

Al-Qabbanji proposes "a modifiable religious ruling based on *fiqh al-maqasid*, or the Jurisprudence of the Meaning."[30] According to this innovation, "jurisprudence should address the meaning conveyed by the revelation, rather than adhere blindly to its literal wording, with no regard for reality or reason."[31] Al-Qabbanji has proposed viewing the Qur'an as divinely inspired *but not divinely dictated*, a break with current orthodoxy. Al-Qabbanji believes that "the Qur'an was created by the Prophet Muhammad, but was driven by Allah."[32] Al-Qabbanji argues that structural reforms are needed within Islam to prevent its stagnation: "If we want Islam to be eternal even though reality is mobile, then Islam must also be mobile. It cannot stagnate. The scholars in the religious institutions view Islam as stagnant teachings."[33]

Another reformer worth noting is Iyad Jamal al-Din, an Iraqi cleric. Though he is a Shiite, al-Din has argued *against* political rule by clerics as occurs in Iran, and for separation of mosque and state, and has faced numerous threats for taking these positions. Al-Din rejects the imposition of sharia and favors civil laws in a civil state in order to guarantee full freedom of conscience to each individual citizen:

> I say that either we follow the *fiqh* [Islamic religious law], in which case ISIS is more or less right, or else we follow man-made, civil enlightened law, according to which the Yazidis are citizens just like Shiite and Sunni Muslims. We must make a decision whether to follow man-made civil law, legislated by the Iraqi parliament, or whether to follow the fatwas issued by Islamic jurisprudence. We must not embellish things and say that Islam is a religion of compassion, peace and rose water, and that everything is fine.[34]

Al-Din has defended the religious diversity of Iraq and has rebuked IS on theological grounds for imposing its religious views on nonbelievers. He has described the first article in most Islamic constitutions, which declares the state to be an Islamic state, as "a catastrophe." He argues that "religion is for human beings, not the state."[35]

Ibrahim al-Buleihi, a former member of the Saudi Shura council who has held a number of government posts, has publicly stated that the Arab world needs a fundamental cultural change to empower *the individual* and make possible independent thinking.[36] Al-Buleihi rejects the groupthink and tendency toward public conformity that has constrained independent thinking in the Islamic world. Independent thinking, outside of the shackles of orthodoxy, is necessary for a civilization to flourish.

Similarly, Dhiyaa al-Musawi, a Bahraini Shia cleric, thinker, and writer, has called "for a cultural Intifada in the Arab world, in order to sweep away the superstitions that dwell in the Arab and Islamic mind."[37]

Reformers and the West

Just as critics of communism during the Cold War came from a variety of backgrounds and disagreed on much, today's critics of Islam unreformed are not in agreement on all issues. Al-Qabbanji, for example, has expressed strong criticism of U.S. and Israeli foreign policy. Other reformers, such as al-Ansari, are generally pro-American in inclination.

Those Muslim reformers who propose breaking with Islamic orthodoxy to empower the individual, who want to create a civil state under civil laws, who view the Qur'an as a document created by men, and who support critically analyzing the Qur'an and the hadith—these individuals are ultimately allies of human freedom though they may differ with Westerners on matters of public policy. These men risk imprisonment and even death in order to reform Islam from within and change its core doctrines. They merit our support—though they are unlikely to agree with Westerners on every matter of foreign policy.

I do not believe, as some people do, in the innate "backwardness" of Arabs or of Muslims, or for that matter of Africans or Somalis. I do not believe Islamic orthodoxy is "ingrained" in the nature of Muslims. I do not believe the Islamic world is doomed to a perpetual cycle of violence, whoever succeeds in reaching the levers of power. And I do not believe that Islamic clerics—guardians of orthodoxy—are powerful enough to stop a groundswell of dissatisfaction with the existing state of affairs.

I am a universalist. I believe that each human being possesses the power of reason as well as conscience. That includes all Muslims as individuals. At present, some Muslims ignore their consciences, and join groups such as Boko Haram or IS, obeying textual prescriptions and religious dogma.

But their crimes against human reason and against human conscience committed in the name of Islam and sharia are already forcing a reexamination of Islamic scripture, doctrine, and law. This process cannot be stopped, no matter how much violence is used against would-be reformers. Ultimately, I believe it is human reason and human conscience that will prevail.

It is the duty of the Western world to provide assistance and, where necessary, security to those dissidents and reformers who are carrying out this formidable task. Dissidents have many disagreements among them: what unites them is a concern that Islam unreformed provides neither a viable ethical framework nor a strong connection to the Divine, to the realm beyond. To repeat the words of al-Din, "We must not embellish things and say that Islam is a religion of compassion, peace and rose water, and that everything is fine." It is not. But the fact that such words can be uttered at all is one of the reasons I believe the Muslim Reformation has begun.

· NOTES ·

INTRODUCTION: *On Islam, Three Sets of Muslims*

1. Sarah Fahmy, "Petition: Speak Out Against Honoring Ayaan Hirsi Ali at Brandeis' 2014 Commencement." https://www.change.org/p/brandeis-university-administration-speak-out-against-honoring-ayaan-hirsi-ali-at-brandeis-2014-commencement.
2. Ibid.
3. Brandeis Faculty Letter to President Lawrence Concerning Hirsi Ali, April 6, 2014. Available at https://docs.google.com/document/d/1M0AvrWuc3V0nMFqRDRTkLGpAN7leSZfxo3y1msEyEJM/edit?pli=1.
4. Letter found on the body of Theo van Gogh, 2004. http://vorige.nrc.nl/krant/article1584015.ece.
5. Asra Nomani, "The Honor Brigade," *Washington* Post, January 16, 2015. http://www.washingtonpost.com/opinions/meet-the-honor-brigade-an-organized-campaign-to-silence-critics-of-islam/2015/01/16/0b002e5a-9aaf-11e4-a7ee-526210d665b4_story.html.
6. Soren Seelow, "It's Charlie, Hurry, They're All Dead," *Le Monde*, January 13, 2015. http://www.lemonde.fr/societe/article /2015 /01/13/c-est-charlie-venez-vite-ils-sont-tous-morts_4554839 _3224.html.

7. Norman Cohn, *The Pursuit of the Millennium: Revolutionary Millenarians and Mystical Anarchists of the Middle Ages* (New York: Oxford University Press, 1957).
8. Pew Research Center, "The World's Muslims: Religion, Politics and Society," 2013. http://www.pewforum.org/2013/04/30/the-worlds-muslims-religion-politics-society-overview/.
9. Kevin Sullivan, "Three American Teens, Recruited Online, Are Caught Trying to Join the Islamic State," *Washington Post*, December 8, 2014. http://www.washingtonpost.com/world/national-security/three-american-teens-recruited-online-are-caught- trying-to-join-the-islamic-state/2014/12/08/80 22e6c4-7afb-11e4-84d4-7c896b90abdc_story.html.
10. UN Security Council, 7316th Meeting, November 19, 2014. http://www.un.org/press/en/2014/sc11656.doc.htm. See also Spencer Ackerman, "Foreign Jihadists Flocking to Syria on 'Unprecedented Scale'—UN," *Guardian*, October 30, 2014. http://www.theguardian.com/world/2014/oct/30/foreign-jihadist-iraq-syria-unprecedented-un-isis.
11. *Economist*, "It Ain't Half Hot Here, Mum: Why and How Westerners Go to Fight in Syria and Iraq," August 30, 2014. http://www.economist.com/news/middle-east-and-africa/21614226-why-and-how-westerners-go-fight-syria-and-iraq-it-aint-half-hot-here-mum.
12. Pew Research Center, "The Future of the Global Muslim Population: Projections for 2010–2030," 2011.
13. Pew Research Center, "The World's Muslims: Religion, Politics and Society," 2013.
14. Ibid., "Survey Topline Results": Apostasy (Q92b), Belief in God (Q16), Duty to convert (Q52), Sharia revealed word (Q66), Influence of religious leaders (Q15), Western entertainment (Q26), Polygamy (84b), Honor killings (Q54), Suicide bombings (Q89), Divorce (Q77), Daughter marrying a Christian (Q38). http://www.pewforum.org/files/2013/04/worlds-muslims-religion-politics-society-topline1.pdf.

CHAPTER 1: *The Story of a Heretic*

1. Sohrab Ahmari, "Inside the Mind of the Western Jihadist," *Wall Street Journal*, August 30, 2014. http://www.wsj.com/articles/SB20001424052970203977504580115831289875638.
2. Ibid.
3. Michele McPhee, "Image Shows Dzhokhar Tsarnaev's Last Message Before Arrest," ABC News, April 17, 2014. http://abc news.go.com/Blotter/image-shows-dzhokhar-tsarnaevs-message-arrest/story?id=23335984&page=2.
4. Ahmari, "Inside the Mind of the Western Jihadist."

CHAPTER 2: *Why Has There Been No Muslim Reformation?*

1. Nonie Darwish, "Qaradawi: If They [Muslims] Had Gotten Rid of the Punishment for Apostasy, There Would Be No Islam Today," February 5, 2013. http://www.gatestoneinstitute.org/3572/islam-apostasy-death. Original footage available at https://www.youtube.com/watch?v=tB9UdXAP82o.
2. Pew Research Center, "In 30 Countries, Heads of State Must Belong to a Certain Religion," 2014. http://www.pewresearch.org/fact-tank/2014/07/22/in-30-countries-heads-of-state-must-belong-to-a-certain-religion/.
3. Daniel Philpott, *Revolutions in Sovereignty: How Ideas Shaped Modern International Relations* (Princeton: Princeton University Press, 2001), p. 81.
4. Albert Hourani, *Arabic Thought in the Liberal Age, 1798–1939* (Cambridge: Cambridge University Press, 1983), p. 247.
5. "Hassan al Banna" in *Princeton Readings in Islamist Thought*, edited by Roxanne Euben and Muhammad Qasim Zaman (Princeton: Princeton University Press, 2009), pp. 49–55.
6. Hourani, *Arabic Thought in the Liberal Age*, p. 8.

7. Sahih al-Bukhari, volume 8, book 76, no. 437.
8. Ella Landau-Tasseron, "The 'Cyclical Reform': A Study of the Mujaddid Tradition," *Studia Islamica* 70 (1989): 79–117.
9. David Bonagura, "Faith and Emotion," *The Catholic Thing*, February 6, 2014. http://thecatholicthing.org/2014/02/06/faith-and-emotion/. Accessed December 18, 2014.
10. Elizabeth Flock, "Saudi Blogger's Tweets about Prophet Muhammad Stir Islamists to Call for His Execution," *Washington Post*, February 9, 2012. http://www.washingtonpost.com/blogs/worldviews/post/saudi-bloggers-tweets-about-prophet-muhammad-stir-islamists-to-call-for-his-execution/2012/02/09/gIQATqbc1Q_blog.html.
11. Ibid.
12. Pew Research Institute, "Concerns about Islamic Extremism on the Rise in Middle East," 2014. http://www.pewglobal.org/2014/07/01/concerns-about-islamic-extremism-on-the-rise-in-middle-east/.
13. Raymond Ibrahim, "Egypt's Sisi: Islamic 'Thinking' Is 'Antagonizing the Entire World,'" January 1, 2015. http://www.raymondibrahim.com/from-the-arab-world/egypts-sisi-islamic-thinking-is-antagonizing-the-entire-world/. Emphasis added.
14. Shmuel Sasoni, "Son's Suicide Is Rohani's Dark Secret," *Ynet Middle East*, June 18, 2013. http://www.ynetnews.com/articles/0,7340,L-4393748,00.html.

CHAPTER 3: *Muhammad and the Qur'an*

1. Ernest Gellner, *Muslim Society* (Cambridge: Cambridge University Press, 1981), p. 1.
2. Sahih Muslim, book 19, nos. 4464, 4465, 4466, 4467.
3. Gerhard Bowering, "Muhammad (570–632)," in *The Princeton*

Encyclopedia of Islamic Political Thought, edited by Gerhard Bowering (Princeton: Princeton University Press, 2013), pp. 367–75.

4. Qur'an, Yusufali translation. University of Southern California Center for Muslim-Jewish Engagement. http://www.usc.edu/org/cmje/religious-texts/quran/verses/033-qmt.php.
5. Philip Carl Salzman, "The Middle East's Tribal DNA." *Middle East Quarterly* (2008): 23–33.
6. Philip Carl Salzman, *Culture and Conflict in the Middle East* (Amherst: Humanity Books, 2008).
7. Gerhard Bowering, a professor of Islamic studies at Yale, summarizes the transition from Arab tribes to Muslim supertribe as follows: "For the first time in history, the tribal energy of the Arab clansmen, spent in the past on nomadic raids or tribal blood feuds, became directed towards the common goal of building a coordinated polity. This polity was to be driven by jihad." "Muhammad (570–632)," in *The Princeton Encyclopedia of Islamic Political Thought*.
8. Patricia Crone, "Traditional Political Thought," in ibid., p. 559.
9. See Sahih Bukhari, book 53 (Khumus) and book 59 (Al-Maghaazi). University of Southern California Center for Muslim-Jewish Engagement. http://www.usc.edu/org/cmje/religious-texts/hadith/bukhari/.
10. Antony Black, *The History of Islamic Political Thought* (Edinburgh: Edinburgh University Press, 2001).
11. Patricia Crone, *God's Rule: Government and Islam* (New York: Columbia University Press, 2004), p. 10.
12. For an analysis of determinism in Islamic history, see Suleiman Ali Mourad, "Free Will and Predestination," in *The Islamic World*, edited by Andrew Rippin (New York: Routledge, 2008), pp. 179–90.
13. Ibid.
14. Tawfik Hamid, "Does Moderate Islam Exist?" *Jerusalem Post*, September 14, 2014. http://www.jpost.com/Experts/Does-moderate-Islam-exist-375316.

15. Ibid.
16. Ibid.
17. Yahya Michot, "Revelation," in *The Cambridge Companion to the Quran*, edited by Jane Dammen McAuliffe (Cambridge: Cambridge University Press, 2006), pp. 180–96.
18. Harald Motzki, "Alternative Accounts of the Quran's Formation," in ibid., p. 60.
19. John Wansbrough, *Quranic Studies: Sources and Methods of Scriptural Interpretation* (Oxford: Oxford University Press, 1977), and *The Sectarian Milieu: Content and Composition of Islamic Salvation History* (Oxford: Oxford University Press, 1978).
20. Fred Donner, "The Historical Context," in *The Cambridge Companion to the Qur'an*, pp. 23–40.
21. Claude Gilliot, "Creation of a Fixed Text," in ibid., pp. 41–58.
22. Arthur Jeffery, "Abu 'Ubaid on the Verses Missing from the Quran," in *The Origins of the Quran: Classic Essays on Islam's Holy Book*, edited by Ibn Warraq (Amherst: Prometheus Books, 1998), pp. 150–54.
23. Toby Lester, "What Is the Quran?" *Atlantic*, January 1, 1999. http://www.theatlantic.com/magazine/archive/1999/01/what-is-the-Quran/304024/.
24. Motzki, "Alternative Accounts of the Quran's Formation," pp. 59–75.
25. Michael Cook, "The Collection of the Quran," in *The Quran: A Short Introduction* (Oxford: Oxford University Press, 2000), pp. 119–26.
26. Malise Ruthven, *Islam in the World* (Oxford: Oxford University Press, 2006), p. 81, emphases added.
27. Ibid.
28. Ibn Warraq, "Introduction," in *Which Quran? Variants, Manuscripts, Linguistics*, edited by Ibn Warraq (Amherst: Prometheus Books, 2011), p. 44. Warraq refers to Abul A'la Mawdudi, *Towards Understanding Islam* (Gary, IN: International Islamic Federation of Student Organizations, 1970).
29. Raymond Ibrahim, "How Taqiyya Alters Islam's Rules of War"

Middle East Quarterly (2010): pp. 3–13. http://www.meforum.org/2538/taqiyya-islam-rules-of-war.

30. David Bukay, "Peace or Jihad? Abrogation in Islam," *Middle East Quarterly*, 2007, pp. 3–11.
31. Raymond Ibrahim, "Ten Ways Islam and the Mafia Are Similar," 2014. http://www.raymondibrahim.com/islam/ten-ways-the-mafia-and-islam-are-similar/.
32. Bukay, "Peace or Jihad? Abrogation in Islam."
33. Andrew Higgins, "The Lost Archive: Missing for a Half Century, a Cache of Photos Spurs Sensitive Research on Islam's Holy Text," *Wall Street Journal*, January 12, 2008. http://online.wsj.com/articles/SB120008793352784631.
34. Ibid.
35. Michael Cook, *The Quran: A Short Introduction* (Oxford: Oxford University Press, 2000), pp. 77, 80, 95, 127.
36. Ibid., p. 79.
37. David Cook, *Understanding Jihad* (Los Angeles: University of California Press, 2005), p. 43.
38. Ibid., p. 32.
39. Ibid., p. 42.
40. Mariam Karouny, "Apocalyptic Prophecies Drive Both Sides to Syrian Battle for End of Time," Reuters, April 1, 2014. http://www.reuters.com/article/2014/04/01/us-syria-crisis-prophecy-insight-idUSBREA3013420140401.
41. Ibid.
42. Ali Khan and Hisham Ramadan, *Contemporary Ijtihad: Limits and Controversies* (Edinburgh: Edinburgh University Press, 2011), p. 36.
43. Christina Phelps Harris, *Nationalism and Revolution in Egypt* (New York: Hyperion Press, 1981 [1964]), p. 111.
44. Jason Burke, "Taliban Prepare for Civilian Rule," *Independent*, August 21, 1998. http://www.independent.co.uk/news/taliban-prepare-for-civilian-rule-1173015.html.
45. Mahmoud Mohamed Taha, *The Second Message of Islam* (Syracuse: Syracuse University Press, 1987).

CHAPTER 4: *Those Who Love Death*

1. Kevin Sullivan, "Three American Teens, Recruited Online, Are Caught Trying to Join the Islamic State," *Washington Post*, December 8, 2014. http://www.washington post.com/world/national-security/three-american-teens-recruited-online-are-caught-trying-to-join-the-islamic-state/2014/12/08/8022 e6c4-7afb-11e4-84d4-7c896b90abdc_story.html.
2. Ibid.
3. Ibid.
4. Asma Afsaruddin, "Martyrdom," in *The Princeton Encyclopedia of Islamic Political Thought*, p. 329.
5. Imam Al-Ghazzali, *Ihya Ulum-id-Din* (Karachi: Darul-Ishaat), vol. 4, p. 428.
6. Jane Idleman Smith and Yvonne Yazbeck Haddad, "The Special Case of Women and Children in the Afterlife," in *The Islamic Understanding of Death and Resurrection* (Albany: SUNY Press, 1981), pp. 157–82.
7. Sermon by Sheikh Muhammad Hassan. 13:34. https://www.youtube.com/watch?v=7i92a3oKkGk.
8. Discussion based on Terence Penelhum, "Christianity," in *Life After Death in World Religions*, edited by Harold Coward (Maryknoll: Orbis, 1997), pp. 31–47.
9. Thomas Hegghammer, "Suicide," in *The Princeton Encyclopedia of Islamic Political Thought*, pp. 530–31.
10. Sullivan, "Three American Teens."
11. John Estherbrook, "Salaries for Suicide Bombers," CBS News, April 3, 2002. http://www.cbsnews.com/news/salaries-for-suicide-bombers/.
12. MEMRI, "Gaza Lecturer Subhi Al-Yazji: Suicide Bombers Are Motivated by Islamic Faith, Not Financial Need or Brainwashing," 2014. http://www.memri.org/clip_transcript/en/4318.htm.
13. Itamar Marcus, "Islamic Law and Terror in Palestinian Authority

Ideology," Palestinian Media Watch, 2002. http://www.palwatch.org/main.aspx?fi=155&doc_id=2321.

14. Raphael Israeli, *Islamikaze: Manifestations of Islamic Martyrology* (New York: Routledge, 2003), p. 216.
15. Al-Risala, July 7, 2001.
16. Palestinian Media Watch, January 1, 2006.
17. Palestinian Media Watch, "Success of Shada Promotion," 2006. http://palwatch.org/main.aspx?fi=635&fld_id=635&doc_id=1109.
18. Palestinian Media Watch, "Martyrs Rewarded with 72 Virgins," 2004. http://palwatch.org/main.aspx?fi=565.
19. MEMRI, "Ten-Year-Old Yemeni Recites Poetry about the Liberation of Jerusalem," 2010. http://www.memritv.org/clip_transcript/en/2723.htm.
20. Drew Hinshaw, "Children Enlist in African Religious Battles," *Wall Street Journal*, July 1, 2014.
21. http://www.thedailybeast.com/articles/2014/08/06/the-isis-online-campaign-luring-western-girls-to-jihad.html.
22. Shamim Siddiqi, *Methodology of Dawah Il Allah in American Perspective* (Brentwood: International Graphic, 1989), chapter 3, p. 33.
23. James Burke, *The Day the Universe Changed* (New York: Hachette, 1985), p. 38.
24. Albert Hourani, *Arabic Thought in the Liberal Age, 1798–1939* (Cambridge: Cambridge University Press, 1983), pp. 41–42.
25. Maribel Fierro, "Heresy and Innovation," in *The Princeton Encyclopedia of Islamic Political Thought*, pp. 218–19.
26. Televised interview of Zakir Naik by Shahid Masood on ARY Digital. Available at https://www.youtube.com/watch?v=6jYUL7eBdHg.
27. "Open Letter to Al-Baghdadi and to the Fighters and Followers of the Self-Declared 'Islamic State,'" 2014. http://www.lettertobaghdadi.com/.
28. Timur Kuran, *The Long Divergence: How Islamic Law Held Back the Middle East* (Princeton: Princeton University Press, 2011).

CHAPTER 5: *Shackled by Sharia*

1. Harriet Alexander, "Meriam Ibrahim 'Should Be Executed,' Her Brother Says," *Telegraph*, June 5, 2014. http://www.telegraph.co.uk/news/worldnews/africaandindianocean/sudan/10877279/Meriam-Ibrahim-should-be-executed-her-brother-says.html.
2. Cited in Ernest Gellner, *Muslim Society* (Cabridge: Cambridge University Press, 1988), p. 1.
3. Patricia Crone, *God's Rule: Government and Islam* (New York: Columbia University Press, 2004), p. 287.
4. Gellner, *Muslim Society*, p. 1.
5. Dan Diner, *Lost in the Sacred: Why the Muslim World Stood Still* (Princeton: Princeton University Press, 2009).
6. Ibid.
7. http://www.cnn.com/2015/01/21/middleeast/saudi-beheading-video/.
8. BBC, "What Are Pakistan's Blasphemy Laws?" November 6, 2014. http://www.bbc.com/news/world-south-asia-12621225.
9. Nurdin Hasan, "Aceh Government Removes Stoning Sentence from Draft Bylaw," *Jakarta Post*, March 12, 2013. http://thejakartaglobe.beritasatu.com/news/aceh-government-removes-stoning-sentence-from-draft-bylaw/.
10. Richard Edwards, "Sharia Courts Operating in Britain," *Telegraph*, September 14, 2008. http://www.telegraph.co.uk/news/uknews/2957428/Sharia-law-courts-operating-in-Britain.html.
11. Maryam Namazie, "What Isn't Wrong with Shariah Law?" *Guardian*, July 5, 2010. http://www.theguardian.com/law/2010/jul/05/sharia-law-religious-courts.
12. Ruud Koopmans, "Fundamentalism and Out-Group Hostility: Immigrants and Christian Natives in Western Europe," WZB Berlin, 2013. http://www.wzb.eu/sites/default/files/u6/koopmans_englisch_ed.pdf.
13. Alex Schmid, "Violent and Non-violent Extremism: Two Sides

of the Same Coin?" ICCT Research Paper, The Hague, 2014, p. 8.

14. Ahmad ibn Nagil al-Misri, *Reliance of the Traveller: A Classical Manual of Islamic Sacred Law* (Beltsville: Amana, 1997), F 5.3.
15. Ibid., M 10.12, p. 541.
16. Ibid., M 3.13, M 3.15.
17. Richard Antoun, "On the Modesty of Women in Arab Muslim Villages: A Study in the Accommodation of Traditions," *American Anthropologist* 70 (4): 671–97.
18. Phyllis Chesler, "Are Honor Killings Simply Domestic Violence?" *Middle East Quarterly*, 2009, pp. 61–69.
19. Aymenn Jawad Al-Tamimi, "The Problem of Honor Killings," *Foreign Policy Journal*, September 2010. http://www.foreignpolicyjournal.com/2010/09/13/the-problem-of-honor-killings/.
20. Yotam Feldner, "'Honor' Murders—Why the Perps Get Off Easy," *Middle East Quarterly*, 2000, pp. 41–50. http://www.meforum.org/50/honor-murders-why-the-perps-get-off-easy. Emphases added.
21. MEMRI, "Egyptian Cleric Sa'd Arafat: Islam Permits Wife Beating Only When She Refuses to Have Sex with Her Husband, 2010." http://www.memritv.org/clip_transcript/en/2600.htm.
22. Brian Whitaker, "From Discrimination to Death—Being Gay in Iran," *Guardian*, December 15, 2010. http://www.theguardian.com/commentisfree/2010/dec/15/gay-iran-mahmoud-ahmadinejad.
23. IRQO, *The Violations of the Economic, Social, and Cultural Rights of Lesbian, Gay, Bisexual, and Transgender (LGBT) Persons in the Islamic Republic of Iran*, 2012. http://www2.ohchr.org/English/bodies/cescr/docs/ngos/JointHeartlandAlliance_IRQO_IHRC_Iran_CESCR50.pdf. See also Vanessa Barford, "Iran's 'Diagnosed Transsexuals,'" BBC, February 25, 2008. http://news.bbc.co.uk/2/hi/7259057.stm.
24. Pew Research Forum, "The World's Muslims: Religion, Politics

and Society," 2013. http://www.pewforum.org/2013/04/30/the-worlds-muslims-religion-politics-society-overview/.

25. Daniel Howden, "'Don't Kill Me,' She Screamed. Then They Stoned Her to Death," *Independent*, November 9, 2008. http://www.independent.co.uk/news/world/africa/dont-kill-me-she-screamed-then-they-stoned-her-to-death-1003462.html.
26. Betty Friedan, *The Feminine Mystique* (New York: Norton, 1997), p. 144.

CHAPTER 6: *Social Control Begins at Home*

1. Michael Cook, *Forbidding Wrong in Islam: A Short Introduction* (Cambridge: Cambridge University Press, 2003), p. 147.
2. Patricia Crone, *God's Rule: Government in Islam* (New York: Columbia University Press, 2004), pp. 300–301.
3. Ben Quinn, "'Muslim Patrol' Vigilante Pleads Guilty to Assaults and Threats," *Guardian*, October 13, 2013. http://www.theguardian.com/uk-news/2013/oct/18/muslim-patrol-vigilante-guilty-assault.
4. "Locals Concerned as 'Sharia Police' Patrol Streets of German City," *Deutsche Welle*, 2014. http://www.dw.de/locals-concerned-as-sharia-police-patrol-streets-of-german-city/a-17904887.
5. Pakistan Human Rights Commission, *State of Human Rights in 2013.* www.hrcp-web.org/hrcpweb/report14/AR2013.pdf.
6. Terrence McCoy, "In Pakistan, 1,000 Women Die in 'Honor Killings' Annually. Why Is This Happening?" *Washington Post*, May 28, 2014. http://www.washingtonpost.com/news/morning-mix/wp/2014/05/28/in-pakistan-honor-killings-claim-1000-womens-lives-annually-why-is-this-still-happening/.
7. Aymenn Jawad, Al-Tamimi, "The Problem of Honor Killings," *Foreign Policy Journal*, September 2010. http://www.foreignpolicy.com/2010/09/13/the-problem-of-honor-killings
8. Dawood Azami, "Controversy of Apostasy in Afghanistan,"

BBC, January 14, 2014. http://www.bbc.com/news/world-asia-25732919.

9. Jeffrey Goldberg, "The Modern King in the Arab Spring," *Atlantic*, April 2013. http://www.theatlantic.com/magazine/archive/2013/04/monarch-in-the-middle/309270/?single_page=true.
10. Ibid.
11. Cook, *Forbidding Wrong in Islam*, pp. 114–15, 122.
12. Patricia Crone, "Traditional Political Thought," in *The Princeton Encyclopedia of Islamic Political Thought*, pp. 554–60.
13. Kathy Gilsinan, "The ISIS Crackdown on Women, by Women," *Atlantic*, July 25, 2014. http://www.theatlantic.com/international/archive/2014/07/the-women-of-isis/375047/.
14. Nadya Labi, "An American Honor Killing: One Victim's Story," *Time*, February 25, 2011. http://content.time.com/time/nation/article/0,8599,2055445,00.html.
15. "Brother of Slain Girls Defends Father at Vigil," NBC News, March 9, 2008. http://www.nbc5i.com/newsarchive/15546408/detail.html.
16. Oren Yaniv, "Pakistani Man Gets 18 Years to Life for Beating Wife to Death After She Made Lentils for Dinner," July 9, 2014. http://www.nydailynews.com/new-york/nyc-crime/pakistani-man-18-years-life-beating-wife-death-made-lentils-dinner-article-1.1860459.
17. "Derby Gay Death Call Leaflet Was 'Muslim Duty,'" BBC, January 12, 2012. http://www.bbc.com/news/uk-england-derbyshire-16581758.
18. Kunal Dutta, "ISIS Suicide Bomber from Derby Thought to Have Killed Eight in Iraq 'Could Have Been Brainwashed,'" *Independent*, November 9, 2014. http://www.independent.co.uk/news/world/middle-east/isis-suicide-bomber-from-derby-kills-eight-in-iraq-9849307.html.
19. James Harkin, "Inside the Mind of a British Suicide Bomber," *Newsweek*, November 21, 2014. http://www.newsweek.com/2014/11/21/inside-frenzied-mind-british-suicide-bomber-283634.html.

20. "Muslim Radio Station Fined for Saying People Should Be Tortured," *Daily Telegraph*, November 23, 2012. http://www.telegraph.co.uk/news/religion/9698967/Muslim-radio-station-fined-for-saying-gay-people-should-be-tortured.html.
21. Ibid.

CHAPTER 7: *Jihad*

1. Capital Bay News, "Lee Rigby Trial Updates," 2013. http://www.capitalbay.com/news/432534-live-lee-rigby-trial-updates-as-michael-adebolajo-and-michael-adebowale-stand-accused-of-woolwich-soldier-murder.html.
2. "Text from Dzokhar Tsarnaev's Note Written in Watertown Boat," *Boston Globe*, May 22, 2014. http://www.bostonglobe.com/metro/2014/05/22/text-from-dzhokhar-tsarnaev-note-left-watertown-boat/KnRIeqqr95rJQbAbfnj5EP/story.html.
3. Ibid.
4. Sebastian L. v. Gorka, "The Enemy Threat Doctrine of Al Qaeda: Taking the War to the Heart of Our Foe," in *Fighting the Ideological War: Winning Strategies from Communism to Islamism*, edited by Katherine C. Gorka and Patrick Sookhdeo (McLean: Isaac Publishing, 2012), pp. 198–201.
5. David Cook, *Understanding Jihad* (Los Angeles: University of California Press, 2005), pp. 32–33.
6. Rajia Aboulkeir, "Meet Islam Yaken, a Cosmopolitan Egyptian Who Turned into ISIS Fighter," *Al-Arabiya*, August 3, 2014. http://english.alarabiya.net/en/variety/2014/08/03/Meet-Islam-Yaken-a-cosmopolitan-Egyptian-who-turned-into-ISIS-fighter-.html.
7. Hamas, "Boy Vows to Join Father in Martyrs' Paradise," 2009. http://palwatch.org/main.aspx?fi=585&fld_id=633&doc_id=2789.
8. AIVD, *The Transformation of Jihadism in the Netherlands: Swarm*

Dynamics and New Strength (The Hague, 2014). https://www.aivd.nl/english/publications-press/@3139/transformation-0/.

9. Bart Olmer, "Threat of Jihadists Greater Than Ever," *De Telegraaf*, June 30, 2014.
10. Ibid.
11. Ibid.
12. Pew Research Institute, "Muslim Americans: Middle Class and Mostly Mainstream," 2007, p. 6.
13. Pew Research Institute, "Muslim Americans: No Signs of Growth in Alienation or Support for Extremism," 2011, p. 4.
14. Dominic Evans, "Exiled Cleric Who Taught UK Knifeman Praises Courage," Reuters, May 24, 2013. http://www.reuters.com/article/2013/05/24/us-britain-killing-bakri-idUSBRE94N0D920130524.
15. Patricia Crone, "Traditional Islamic Political Thought," in *The Princeton Encyclopedia of Islamic Political Thought*.
16. Human Rights Watch, "Nigeria: Boko Haram Kills 2,053 Civilians in 6 Months," July 15, 2014. http://www.hrw.org/news/2014/07/15/nigeria-boko-haram-kills-2053-civilians-6-months.
17. UNHCR. 2015 UNHCR Country Operations Profile. http://www.unhcr.org/pages/4e43cb466.html.
18. Pew Research Center, "Global Christianity: A Report on the Size and Distribution of the World's Christian Population," 2011, p. 64.
19. André Aciman, "After Egypt's Revolution, Christians Are Living in Fear," *New York Times*, November 19, 2011. http://www.nytimes.com/2011/11/20/opinion/sunday/after-egypts-revolution-christians-are-living-in-fear.html.
20. Richard Spencer, "Egypt's Coptic Christians Fleeing Country After Islamist Takeover," *Telegraph*, January 13, 2013. http://www.telegraph.co.uk/news/worldnews/africaandindianocean/egypt/9798777/Egypts-Coptic-Christians-fleeing-country-after-Islamist-takeover.html.

21. Nina Shea, Paul Marshall, and Lela Gilbert, *Saudi Arabia's Curriculum of Intolerance, with Excerpts from Saudi Ministry of Education Textbooks for Islamic Studies* (Washington, D.C.: Hudson Institute Center for Religious Freedom and the Institute for Gulf Affairs, 2008), pp. 7, 43. http://www.hudson.org/content/researchattachments/attachment/656/saudi_textbooks_final.pdf.
22. "UK Jihad Fighter in Downing Street Flag Threat," *Scotsman*, July 5, 2014. http://www.scotsman.com/mobile/news/uk/uk-jihad-fighter-in-downing-street-flag-threat-1-3467362.
23. Mark Townsend, "British Muslims' Right to Fight in Syria Backed by an Ex-Adviser on Radicalization," *Guardian*, June 28, 2014. http://www.theguardian.com/uk-news/2014/jun/28/british-jidahis-syria-defended.
24. Nadim Roberts, "The Life of a Jihadi Wife: Why One Canadian Woman Joined ISIS's Islamic State," CBC, July 7, 2014. http://www.cbc.ca/news/world/the-life-of-a-jihadi-wife-why-one-canadian-woman-joined-isis-s-islamic-state-1.2696385.
25. Press Association, "British Jihadist Warns of 'Black Flag of Islam' over Downing Street," *Guardian*, July 4, 2014. http://www.theguardian.com/uk-news/2014/jul/04/british-jihadi-black-flag-islam-downing-street.
26. Ibid.
27. Jessica Stern, "Mind over Martyr: How to Deradicalize Islamic Extremists," *Foreign Affairs*, January/February 2010.
28. Elizabeth Dickinson, "Rise of IS Elicits Soul Searching in Arab Gulf, a Source of Funds and Fighters," *Christian Science Monitor*, October 13, 2014. http://www.csmonitor.com/World/Middle-East/2014/1013/Rise-of-IS-elicits-soul-searching-in-Arab-Gulf-a-source-of-funds-and-fighters.
29. Staff, "British Jihadists Urge Their 'Brothers' to Join War," *Times of Israel*, June 21, 2014. http://www.timesofisrael.com/british-citizens-urge-their-brothers-to-join-jihad/.
30. Helen Davidson, "ISIS Instructs Followers to Kill Australians and Other 'Disbelievers,'" *Guardian*, September 23, 2014. http://

www.theguardian.com/world/2014/sep/23/islamic-state-followers-urged-to-launch-attacks-against-australians.

31. See Cook, *Understanding Jihad*, and David Cook, *Martyrdom in Islam* (Cambridge: Cambridge University Press, 2007).

CHAPTER 8: *The Twilight of Tolerance*

1. Adam Wolfson, *Persecution or Toleration: An Explication of the Locke-Proast Quarrel, 1689–1704* (Lanham, MD: Lexington Books, 2010).
2. John Locke, *The Second Treatise of Government and a Letter Concerning Toleration* (Mineola, NY: Dover Publications, 2002).
3. Patrick Kingsley, "80 Sexual Assaults in One Day—the Other Story of Tahrir Square," *Guardian*, July 5, 2013. http://www.theguardian.com/world/2013/jul/05/egypt-women-rape-sexual-assault-tahrir-square.
4. UNICEF, *Female Genital Mutilation/Cutting: A Statistical Overview and Exploration of the Dynamics of Change*, 2013. http://www.unicef.org/publications/index_69875.html.
5. Ali Khan and Hisham Ramadan, *Contemporary Ijtihad: Limits and Controversies* (Edinburgh: Edinburgh University Press, 2011), p. 59.
6. Maribel Fierro, "Heresy and Innovation," in *The Princeton Encyclopedia of Islamic Political Thought* (Princeton: Princeton University Press, 2013), pp. 218–19.
7. Eiynah, "An Open Letter to Ben Affleck," *Pakistan Today*, October 25, 2014. http://www.pakistantoday.com.pk/2014/10/25/comment/an-open-letter-to-ben-affleck/.
8. Michael Warner, "Origins of the Congress for Cultural Freedom," *Studies in Intelligence* 38, no. 5 (1995). See also Peter Coleman, *The Liberal Conspiracy: The Congress for Cultural Freedom and the Struggle for the Mind of Postwar Europe* (New York: Free Press, 1989).

9. Hilton Kramer, "What Was the Congress for Cultural Freedom?" *New Criterion*, 1990. http://www.newcriterion.com/articles.cfm/What-was-the-Congress-for-Cultural-Freedom—5597.
10. Angel Rabasa, Cheryl Bernard, Lowell Schwartz, and Peter Sickle, *Building Moderate Muslim Networks* (Arlington: RAND Corporation, 2007), pp. 17–18. http://www.rand.org/pubs/monographs/MG574.html.
11. Frances Saunders, *The Cultural Cold War: The CIA and the World of Arts and Letters* (New York: Free Press, 1999), p. 89.
12. Barton Gellman and Greg Miller, "'Black Budget' Summary Details U.S. Spy Network's Successes, Failures and Objectives," *Washington Post*, August 29, 2013. http://www.washingtonpost.com/world/national-security/black-budget-summary-details-us-spy-networks-successes-failures-and-objectives/2013/08/29/7e57bb78-10ab-11e3-8cdd-bcdc09410972_story.html.
13. Joseph Stiglitz and Linda Bilmes, *The Three Trillion Dollar War: The True Cost of the Iraq Conflict* (New York: W. W. Norton, 2008); Joseph Stiglitz, "The Price of 9/11," *Project Syndicate*, 2011. http://www.project-syndicate.org/commentary/the-price-of-9-11.

CONCLUSION: *The Muslim Reformation*

1. Quoted in Thomas Friedman, "How ISIS Drives Muslims from Islam," *New York Times*, December 6, 2014.
2. Malala Yousafzai, "Malala Yousafzai: 'Our Books and Our Pens Are the Most Powerful Weapons,' Address to the United Nations," *Guardian*, July 12, 2013. http://www.theguardian.com/commentisfree/2013/jul/12/malala-yousafzai-united-nations-education-speech-text.
3. Yousef Al-Otaiba, "The Moderate Middle East Must Act," *Wall Street Journal*, September 9, 2014. http://www.wsj.com/articles/

yousef-al-otaiba-the-moderate-middle-east-must-act-141 0304537.

4. Ibid. Emphasis added.
5. See Muhammad Abu Samra, "Liberal Critics, 'Ulama' and the Debate on Islam in the Contemporary World," in *Guardians of Faith in Modern Times: 'Ulama in the Middle East*, edited by Meir Hatina (Leiden: Brill, 2008), pp. 265–91.
6. Geneive Abdo, *No God but God: Egypt and the Triumph of Islam* (Oxford: Oxford University Press, 2000), p. 68.
7. Ibid.
8. S. S. Hasan, *Christians versus Muslims in Modern Egypt: The Century-Long Struggle for Coptic Equality* (Oxford: Oxford University Press, 2003), pp. 176–77.
9. Abdo, *No God but God*.
10. Yunis Qandil, "Euro-Islamists and the Struggle for Dominance within Islam," in *The Other Muslims: Moderate and Secular*, edited by Zeyno Baran (New York: Palgrave Macmillan, 2010), pp. 33–55; Hedieh Mirahmadi, "Navigating Islam in America," in *The Other Muslims: Moderate and Secular*, pp. 17–32.
11. "The Enemies of the Muslims According to the Global Islamic Resistance," in Stephen Ulph, "Islamism and Totalitarianism: The Challenge of Comparison," in *Fighting the Ideological War: Winning Strategies from Communism to Islamism*, edited by Katherine C. Gorka and Patrick Sookhdeo (McLean: Isaac Publishing, 2012), p. 75.
12. Quoted in Crone, *God's Rule*, p. 303.
13. Abul 'Ala' Al-Ma'arri [11th century], *The Epistle of Forgiveness: A Vision of Heaven and Hell*, translated by Geert Jan van Gelder and Gregor Schoeler (New York: New York University Press, 2013).
14. France 24, "Jihadists Behead Statue of Syrian Poet Abul Ala al-Maari," February 14, 2013. http://observers.france24.com/content/20130214-jihadists-behead-statue-syrian-poet-abul-ala-al-maari).
15. Reynold Nicholson, *Studies in Islamic Poetry* (Cambridge: Cambridge University Press, 1969).

APPENDIX: *Muslim Dissidents and Reformers*

1. See Ida Lichter, *Muslim Women Reformers: Inspiring Voices Against Oppression* (Amherst: Prometheus Books, 2009); Zeyno Baran, *The Other Muslims: Moderate and Secular* (New York: Palgrave Macmillan, 2010).
2. Zuhdi Jasser, "Americanism vs. Islamism," in Baran, *The Other Muslims: Moderate and Secular*, pp. 175–91.
3. Akbar Ahmed, *Journey into America: The Challenge of Islam* (Washington, DC: Brookings Institution Press, 2010), pp. 238–40.
4. Saleem Ahmed, *Islam: A Religion of Peace?* (Honolulu: Moving Pen Publishers, 2009).
5. Baran, *The Other Muslims: Moderate and Secular.*
6. Yunis Qandil, "Euro-Islamists and the Struggle for Dominance within Islam," in Baran, *The Other Muslims: Moderate and Secular*, pp. 33–55.
7. Lichter, *Muslim Women Reformers*, pp. 346–48.
8. Ibid.
9. Ibid.
10. Samia Labidi, *Karim, mon frère: Ex-intégriste et terroriste* ["Karim, my brother: Former fundamentalist and terrorist"] (Paris: Flammarion, 1997).
11. Lichter, *Muslim Women Reformers*, pp. 346–48.
12. Samia Labidi, "Faces of Janus: The Arab-Muslim Community in France and the Battle for Its Future," in Baran, *The Other Muslims: Moderate and Secular*, pp. 107–22.
13. *Der Spiegel*, "German-Turkish Author Seyran Ateş: 'Islam Needs a Sexual Revolution,'" October 13, 2009. http://www.spiegel.de/international/europe/german-turkish-author-seyran-Ateş-islam-needs-a-sexual-revolution-a-654704.html.
14. Ibid.
15. Poggioli 2008.
16. Abou El-Magd, "Egyptian Blogger Gets 4 Years in Prison,"

Washington Post, February 22, 2007. http://www.washingtonpost.com/wp-dyn/content/article/2007/02/22/AR2007022200269_pf.html.

17. MEMRI, "Egyptian Blogger Abdelkareem Suleiman Arrested for Critizing Al-Azhar Sheikhs," December 7, 2006. http://www.memri.org/report/en/0/0/0/0/0/0/1967.htm.
18. Isabel Kershner, "Palestinian Blogger Angers West Bank Muslims," *New York Times*, November 16, 2010. http://www.nytimes.com/2010/11/16/world/europe/16blogger.html?_r=0.
19. Diaa Hadid, Associated Press, December 6, 2010. http://www.thestar.com/news/world/2010/12/06/palestinian_atheist_jailed_for_weeks_apologizes.html.
20. Kershner, "Palestinian Blogger Angers West Bank Muslims."
21. Hadid, Associated Press, December 6, 2010.
22. Luavut Zahid, "Brandeis University: You've Made a Real Booboo," *Pakistan Today*, April 14, 2014. http://www.pakistantoday.com.pk/2014/04/19/comment/brandeis-university-youve-made-a-real-booboo/.
23. Taslima Nasrin, "They Wanted to Kill Me," *Middle East Quarterly*, 2000. http://www.meforum.org/73/taslima-nasrin-they-wanted-to-kill-me.
24. See Hedieh Mirahmadi, "Navigating Islam in America," in Baran, *The Other Muslims: Moderate and Secular*, pp. 17–32; Yunis Qandil, "Euro-Islamists and the Struggle for Dominance within Islam," in Baran, *The Other Muslims: Moderate and Secular*, pp. 33–55.
25. Hanne Obbink, "Muslims Are Not Allowed to Look Away Any Longer," *Trouw*, December 30, 2014. http://www.trouw.nl/tr/nl/4492/Nederland/article/detail/3819986/2014/12/30/Moslims-mogen-niet-langer-wegkijken.dhtml.
26. Ibid.
27. Interview with al-Ansari, Al-Arabiya TV, May 11, 2007. http://www.memri.org/clip_transcript/en/1450.htm.
28. MEMRI, "Qatari Liberal and Former Dean of Islamic Law at

the University of Qatar: Arab Liberals, Secularists Are Facing Jihad," March 17, 2010. http://www.memri.org/report/en/0/0/0/0/0/0/4041.htm

29. Yotam Feldner, "Liberal Iraqi Shi'ite Scholar Sayyed Ahmad Al-Qabbanji Calls for Reason in Islamic Discourse and Jurisprudence," MEMRI 937, 2013. http://www.memri.org/report/en/0/0/0/0/0/0/7015.htm.
30. Ibid.
31. Ibid.
32. Ibid.
33. Ibid.
34. Ayad Jamal al-Din, "A Civil State in Which All Citizens Are Equal in the Eyes of the Law," Middle East Media Research Institute and Al-Iraqiya TV, October 17, 2014. http://www.memritv.org/clip/en/4556.htm.
35. Nimrod Raphaeli, "Sayyed Ayad Jamal al-Din—Liberal Shi'ite Cleric and Foe of Iran," MEMRI, 2010. http://www.memri.org/report/en/0/0/0/0/0/0/3920.htm.
36. Interview with al-Buleihi on Al-Arabiyya, MEMRI, March 30, 2010. http://www.memritv.org/clip_transcript/en/2414.htm.
37. Interview with al-Musawi on Al-Jazeera, May 4, 2010. http://www.memri.org/clip_transcript/en/2471.htm.

· ABOUT THE AUTHOR ·

Globally known, award-winning human rights activist Ayaan Hirsi Ali is the *New York Times* bestselling author of *Infidel*, *Nomad*, and *The Caged Virgin*. Born in Somalia and raised Muslim, she grew up in Africa and Saudi Arabia, before fleeing to the Netherlands in 1992, where she went from cleaning factories to winning a seat in the Dutch Parliament. A prominent speaker, debater, and op-ed writer, she was chosen as one of *Time* magazine's 100 most influential people in the world. She is now a fellow at Harvard University's John F. Kennedy School of Government. Hirsi Ali is the founder of the AHA Foundation.